2019 年度上海财经大学浙江学院发展基金项
读本"（2019JT001）

婺风家训

（汉英双语版）

主　编	曹艳梅	许　倩		
副主编	徐峥晨	李　慧		
编　委	傅毅强	傅燕芳	李　函	梁　妍
	芮空空	王　芳	王永平	韦杏雨
	吴国清	谢　文	徐玉书	叶丽芳
	郑建明	周铖涛	朱江平	
顾　问	顾卫星	徐　卫	赵　珂	朱　颖

东北大学出版社

·沈　阳·

图书在版编目（CIP）数据

婺风家训：汉文、英文 ／ 曹艳梅，许倩主编. —
沈阳：东北大学出版社，2022.5
　　ISBN 978-7-5517-2969-7

　　Ⅰ. ①婺…　Ⅱ. ①曹…　②许…　Ⅲ. ①家庭道德－金
华－汉、英　Ⅳ. ①B823.1

中国版本图书馆 CIP 数据核字（2022）第 075936 号

出 版 者：东北大学出版社
　　　　　地址：沈阳市和平区文化路三号巷 11 号
　　　　　邮编：110819
　　　　　电话：024 - 83683655（总编室）　83687331（营销部）
　　　　　传真：024 - 83687332（总编室）　83680180（营销部）
　　　　　网址：http://www.neupress.com
　　　　　E-mail: neuph@ neupress.com
印 刷 者：沈阳市第二市政建设工程公司印刷厂
发 行 者：东北大学出版社
幅面尺寸：170 mm×240 mm
印　　张：9.5
字　　数：180 千字
出版时间：2022 年 5 月第 1 版
印刷时间：2022 年 5 月第 1 次印刷
策划编辑：孟　颖
责任编辑：刘新宇
责任校对：杨　坤
封面设计：潘正一
责任出版：唐敏志

ISBN 978-7-5517-2969-7　　　　　　　　　定 价：69.00 元

序

国有国法，家有家规，即没有规矩，不成方圆，人们必须按照法律和规矩办事。国家法律规定了公民的权利和义务，家训也同样为子孙后代制定了立身处世、持家治业的矩规。

最能体现中国古代家族管理的当属家训。家训，别称家诫、家诲、家规、家教，是在家族管理发展过程中总结和提炼出来的一种精神，主要用于家族中长辈对后代的训诫与教诲。家训的形成与家族发展壮大密切相关，也受社会制度的约束。中华传统家训内容一般涉及尊祖敬宗、立志明德、修齐治平、劝勤勉学、立身处世等，具体内容也因家族而异。家训传承了一个民族、一个国家的精神追求和价值取向，是中国优秀文化传统的一部分。不管是对个人还是家庭，甚至对整个社会都有非常好的引领示范作用。

党的十八大以来，习近平总书记多次提到家庭建设和家庭教育的重要性。传承优秀家训，努力培育具有时代精神的优良家风，是传承中华优秀传统文化、坚定文化自信的重要途径。在深入贯彻习总书记有关家庭文明建设的重要论述中，中共金华市委、市纪委高度重视家风建设，深入挖掘各区县优秀传统家规家训，淬炼出了一批有代表性的地方家训精品。

金华素有"小邹鲁"之称，人文荟萃、文风鼎盛，历代名人辈出，留下了丰富的持家治业、修身处世的家训文化。2015年5月22日，中央纪委监察部网站首推"中国传统中的家规"专栏，《郑义门：孝义传家九百年》成为开篇之作。《郑氏规范》共有 168 条，内容非常丰富，从礼仪孝悌、理财治家到为人处世，都做出了明确规定。节目播出后，"江南第一家"的传世家训在全国引起强烈反响。2016 年 3 月 7 日，廉洁动漫片《郑义门》登上央视一套，深入挖掘《郑氏规范》中蕴含的优秀传统文化内涵，汲取家规家风中的精华，引领社会新风尚。

金华题材的家训多次荣登"中国传统中的家规"专栏，譬如第二期的《诸葛村：百世传诵〈诫子书〉》、第三十六期《浙江金华胡则：为官一任 造福一方 读书至乐 教子至要》、第四十六期《浙江金华吕祖谦：明理躬行 清慎勤实》。此外，磐安羊氏家训、义乌徐侨家训、婺城何氏三杰家规家训、金东沈氏家训等陆续登上"中国传统中的家规"专栏，为弘扬传统家训增添了金华特色。中国市雕博物馆专门开辟市雕家训馆，铭辞雕板成为家规家训的独特载体，也荣登中共纪委官网，成为常驻展馆。

家规家训精品能以浓墨重彩的方式重新进入老百姓的视线，与中共金华市委、市纪委的大力倡导息息相关。"八婺好家风"建设八大工程、《关于进一步弘扬优秀传统家规家训 推动党员干部家风建设的实施意见》等一系列党政举措将继承和弘扬优秀传统文化、倡导廉洁正气的优良家风作为推进党风廉政建设的重要工作。

优秀的家训有着穿越时空的巨大魅力，富含亘古不变的人生思考和社会哲理。八婺大地上，古训家风世代传颂，警醒后人，生生不息。

金华市博物馆馆长　徐　卫

2021 年 12 月

Preface

A country has its national law, a family has its family rules; that is to say, where there is no rule, there is no standard. People must abide by the rules and regulations. The national law has formulated the rights and duties of the citizens, whereas the family norms have also set up the rules for their offspring to conduct themselves and govern the family.

It is the family norm that can best reflect the management philosophy of ancient Chinese families. Family norm, also named as family regulation, is a kind of spirit developed and refined from the process of family management, mainly used by the seniors as the norms and teachings for the descendants. The formulation of family norms was closely related to the development and prosperity of the family, which was also constrained by social system. Traditional Chinese norms generally cover the following aspects, like respecting ancestors, setting up aspirations, cultivating morals, managing the family and then the state, being diligent and studious, conducting oneself in the society etc. The specific contents vary from clan to clan. The family norm has inherited the values and the spiritual pursuit of a nation, constituting a significant part of the excellent traditional Chinese culture. It has exerted a leading role in cultivating not only an individual, a family, but also the whole nation.

Since the 18th CPC National Congress, President Xi has mentioned the importance of family construction and family education time and again. Inheriting good family norms and cultivating excellent family tradition in modern times are an important route to carry forward excellent traditional Chinese culture and strengthen cultural confidence. To implement President Xi's important remarks on family cultivation in depth, Jinhua Municipal Party Committee and Discipline Inspection Commission attach great importance to the cultivation of family tradition, probing deep into the fine family regulation of each county, extracting a bunch of typical and exquisite local family norms.

Known as "Xiaozoulu", Jinhua is a place with prosperous culture and outstanding people, leaving behind abundant family norms in governing the family and business, cultivating morals and getting on in life. On May 22, 2015, a special column, "The Family Regulation in Chinese Traditions" was released on the website of Ministry of Supervision of CPC Central Commission for Discipline Inspection. *Zhengyimen*: *Filial Piety Passing down Nine-hundred Years* was the first to be broadcast. With 168 articles, the *Zhengs' Family Regulations* boast of rich contents, regulating clearly what to do concerning filial piety, family governance, business operation, as well as personal conducts. After the broadcast of this program, the classic family instruction of the "The Best Family of Jiangnan" has caused a stir nationwide. On March 7, 2016, *Zhengyimen*, an animation TV series on honesty and integrity, was aired on CCTV-1, probing deep into the excellent cultural connotation embedded in *Zhengs' Regulation*, which has set a new trend in the society by drawing the essence from the family instructions.

The family norms with Jinhua origin have appeared on the special column "The Family Regulation in Chinese Traditions" for many times, such as the second issue—*Zhuge Village: Echoes of "Admonition to My Son" in Centuries*; the thirty-sixth issue—*Hu Ze of Jinhua Zhejiang: Being an Official in One Position Benefits the People in One Place, Reading Brings the Utmost Pleasure While Parenting is the Most Important*; the forty-sixth issue—*Lv Zuqian of Jinhua Zhejiang: Distinguishing Between Right and Wrong in Person; Being Clean, Cautious, Diligent and Honest*. In addition, the family norms of the Yangs of Pan'an, Xuqiao of Yiwu, the Hes of Wucheng District, the Shengs of Jindong District etc. were listed on the above-mentioned special column, which has added Jinhua style in carrying forward the traditional family instructions. China Woodcarving Museum has opened a special hall for carved family norms, where wood-boards inscribed with family norms have become a unique carrier. This hall also got listed on the official website of CPC Central Commission for Discipline Inspection, becoming the permanent exhibition on the website.

Why the great pieces of family norms could make it back to common people's life so strikingly is closely related to the enormous efforts advocated by Jinhua Municipal Party Committee and Discipline Inspection Commission. A series of initiatives, like the construction of "Good Family Tradition in Eight Counties of Wuzhou", *Opinions on Further Promoting Excellent Traditional Family Norms for the Construction of Cadres' Family Tradition*, have been adopted to carry forward the excellent traditional culture and advocate clean and upright family tradition, constituting an important part in the construction of party conduct and clean government.

With everlasting magical charm, excellent family admonitions are rich in the eternal testimony of life philosophies. On the eight counties of Jinhua, ancient family traditions are eulogized from generation to generation, warning the offspring, forever.

Xu Wei

Curator of Jinhua Museum

Dec. 2021

目 录
Contents

第一部分 镌刻·家风
Inscription · Family Tradition

在大力发展传统文化的今天，弘扬和继承中华民族传统美德是我们每一个炎黄子孙的责任。由儒家学派提出的"仁义礼智信"（后称"五常"）贯穿于中华伦理的发展中，成为中国价值体系中的最核心因素。中国古代启蒙教材《三字经》中"曰仁义，礼智信。此五常，不容紊。"将五常之道作为一种道德准则，规范青少年的思想和行为。

Nowadays, it is the duty of every Chinese descendant to carry forward and inherit traditional virtues of the Chinese nation when traditional cultures are greatly developing. "Benevolence (Ren), Righteousness (Yi), Propriety (Li), Wisdom (Zhi), Honesty (Xin)" (later abbreviated as "five virtues") proposed by Confucian school running through the development of the Chinese ethics have become the core of Chinese value system. *Three Character Classic*, an ancient initiatory teaching material in China, records that, "if all people can behave in line with these five unchanged rules, namely benevolence, righteousness, propriety, wisdom, honesty, the society will always peaceful. Hence everyone should bear it in mind and never treat it loosely." It regards five virtues as a kind of moral rule to normalize the mind and behavior of teenagers.

优秀传统文化的继承与发展，既要强调返本，又要强调开新。作为国家历史文化名城、国家创新型城市的金华，创造了灿烂辉煌的地域文化，涌现了大批千古传咏的忠良贤士，发生了众多家喻户晓的孝义故事。"信义金华""廉洁金华"……当这些曾经书写于家规家训上的准则规范同我们城市的名字融合在一起时，所彰显的就是金华人代代相传的淳善清廉之风和为人为官之道。而无论古今，这些被谱写在古志、家谱上的人和事亦以不同的载体和形式进入了百姓生活的方方面面。

The inheritance and development of outstanding traditional culture both focus on returning to the classics and keeping creative. Jinhua, as a state-list historical and cultural city and one of the most innovative Chinese cities, has cultivated splendid territory culture. This land is home to a batch of loyal and wise men praised for ages and many famous stories about filial piety and righteousness. The combinations of regulations and rules written on family rules with the name of our city, like "Faithful Jinhua", "Honest Jinhua", mirror the custom of simpleness and righteousness and the manner of being a man and an official, which has been passed down by Jinhua people since ancient times. No matter in old days or in current times, those people and things written on ancient books and genealogy books have integrated with all aspects of public life through different carriers and forms.

一、忠，敬也 Loyalty(Zhong) means reverence

赵孟垒威武不屈［南宋］ Heroic and indomitable Zhao Menglei ［the Southern Song Dynasty］

赵孟垒，合州人。开庆元年进士，为金华尉。临安投降，与侄儿由鉴携太皇太后帛书到益王处，擢升宗正寺簿、监军。收复明州时，战败被俘，宁死不屈，受磔刑而死。

Zhao Menglei was born in Hezhou. He was the successful candidate in the highest imperial examinations, titling junior officer of Jinhua. When Lin'an was captured, he and his nephew, Youjian, along with silk manuscripts of the emperor's grandmother went to King Yi's place, thus got promoted to take charge of Zongzhengsi and the army. When Mingzhou was recovered, he was captured as a prisoner. Indomitable, he was killed by a kind of cruel torture.

二、孝，德之本也 Filial piety(Xiao) acts as the basis of morality

宗祉孝感动天［明］ The filial deed of Zong Zhi moving the heaven ［the Ming Dynasty］

宗祉，字济才，金华人，游府学。父亲生病，宗祉竭力侍奉汤药。父亲去世后，为父守葬三年。母亲陈氏，双目失明，宗祉便每天早上出村汲取溪水为母亲清洗眼睛，并用舌头为母亲舔舐双目，终使母亲重见光明。母亲陈氏寿终正寝，出葬的那天天降大雨，无法安葬。宗祉嚎啕大哭，向天喊叫求助，天忽然就放晴了，安葬结束后又开始下雨。大家都认为这是宗祉孝感动天所致。

Zong Zhi with his courtesy name as Jicai, was born in Jinhua. He studied in county institute. When his father was ill, he tried his best to serve his father with medicine. After his father's death, he kept vigil of his father's tomb for three years. As for his blind mother Mrs Chen, Zong Zhi would wash his mother's eyes with stream water fetched out of the village every morning. And he would lick his mother's eyes, finally making his mother see the light. His mother, Mrs Chen died a natural death. However, on the day that Zong Zhi planned to bury his mother, it rained heavily, making it impossible to bury his mother. Zong Zhi burst out to cry and yelled at the heaven for help. Suddenly, the sky became clear. But it began raining after the burying. People credited this to the filial deed of Zong Zhi having moved the heaven.

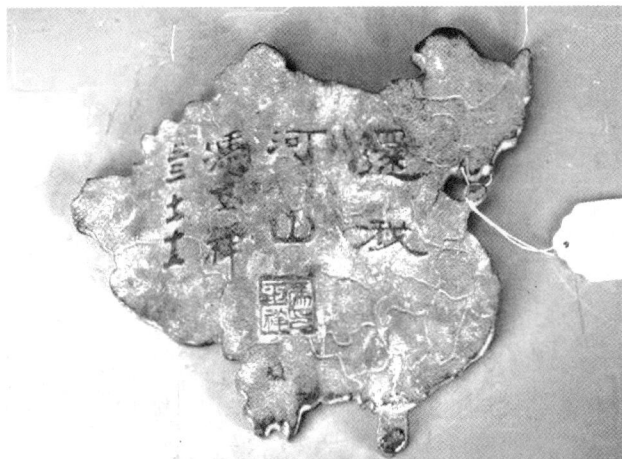

民国冯玉祥题还我河山铜地图 Copper map of returning my land written by Feng Yuxiang of the Republic of China

民国求学不忘救国铭文铜镇纸 Copper paperweight of studying and saving the country of the Republic of China

孟宗哭竹和王祥卧冰　Meng Zong cried to the bamboo and Wang Xiang reclined on the ice

A 面：孟宗哭竹。孟宗年幼的时候父亲离开了人世间。母亲又多病，孟宗不知道该怎么办，跑到竹林里抱着竹子哭诉心中的痛苦。他的孝心感动了大地，大地赐予他一份神秘的礼物，这份礼物治好了他多病的母亲。

B 面：王祥卧冰。王祥继母朱氏生病想吃鲤鱼，但因天寒，河水冰冻，无法捕捉，王祥便赤身卧于冰上，寒冰化开，王祥捕得两条鲤鱼烧给继母吃。

Side A: Meng Zong cried to the bamboo. When Meng Zong was young, his father died. His mother was constantly ill. In face of such living condition, Meng Zong had no idea of how to survive. He ran to the bamboo forest and hugged the bamboo to cry out his inside pain. His filial piety moved the heaven, so the heaven bestowed him a secret gift. This gift cured his ill mother.

Side B: Wang Xiang reclined on the ice. The stepmother of Wang Xiang got sick and wanted to eat carp. However, the cold weather froze the river, making it impossible to catch carp. So Wang Xiang naked himself and reclined on the ice to melt the ice, finally catching two carps to cook for his stepmother.

A   B

民国孟宗哭竹铜花钱 Copper painted money of Meng Zong crying to the bamboo of the Republic of China

三、仁，亲也　Benevolence(Ren) is geniality

倪氏家族乐善好施［明］　The Philanthropic Nis［the Ming Dynasty］

长山石门倪氏家族有一个叫倪永端的人，有乐善好施、仗义疏财的义举，其名被刻于旌善亭。倪永端有个儿子叫倪宗舜，字以仁，天启年间人，也是乐善好施的人，能吟诗，擅教诲。陈氏有一人去世，家贫无以殓葬，宗舜出资为其安葬。倪宗舜的儿子倪梦申，字钟岳，天启年间官至署丞，以勤慎廉察著称。曾经弹劾阉党滥堂役，侵渔者。军需急，倪梦申变卖家产补充军饷，清政府特建清虚输忠坊用以表彰他的义举。倪梦申又捐出白银二千余两，租谷四千余石。先后出米粟赈济饥民，百姓十分信赖他。

There was a man, Ni Yongduan in the Nis of Shimen, Changshan. He made some philanthropic and generous deeds in aiding the needy, so his name was engraved on Jingshan Pavilion. One of Ni Yongduan's sons was Ni Zongshun, with his style name as Yiren, was born in Tianqi period. He was also a philanthropic person who could recite poems and was adept at teaching people. When one of the Chens died but cannot afford a funeral for poverty, Zongshun donated his money to make it. The son of Ni Zongshun, Ni Mengshen with his style name as Zhongyue, worked to Shucheng official in Tianqi period, famous for his studious working and uncorrupted supervision. He once impeached eunuchs for they abused workers and invaded fishers. Due to greatly lacking military supplies, Ni Mengshen sold his property to replenish soldier's pay and provisions. The emperor granted him a memorial gateway inscribed with "清虚输忠(clean-handed working and loyalty to the country)" to

明厚德荣归铭文铜镜 Copper mirror of great-virtue and glory-return inscription in the Ming Dynasty

清厚德荣归铭文铜镜 Copper mirror of great-virtue and glory-return inscription in the Qing Dynasty praise his chivalrous deeds. Ni Mengshen donated over 2,000 liang silvers to borrow grains of over 4,000 shi. He also gave out rice and millet to aid famine refugees, thus making people trust him very much.

四、义，善也 Righteousness(Yi) is kindness

李庶抑强扶弱 [明] **Li Shu curbed the violent and assisted the weak [the Ming Dynasty]**

李庶，字秀卿，福清人，明嘉靖二十一年中进士，任知县，抑强扶弱。有权臣北上，途经兰溪，郡县官员都在道路两旁恭候。李庶说："这是用劳民伤财的行为来保得官位啊！"李庶心里耻于这种做法，到底也没去恭候。嘉靖二十三四年，天大旱，李庶诚心祈雨，着粗布衣，去车步行在烈日下。又平价出售积粟，开仓放粮，用以养活饥民，以此活下来的饥民非常多。李庶死后，其牌位入祀于名宦祠。

Li Shu, with the style name as Xiuqing, was born in Fuqing. He achieved the rank of the Scholar in the 21st years of Jiajing, the Ming Dynasty, and worked as a magistrate of a county. He curbed the violent and assisted the weak. When a powerful official passed Lanxi during the trip to the north, the officials in the county all stood at the sides of the road to welcome. Li Shu said, "they are holding their positions by doing such manpower and money waste." In his heart, he was ashamed of such act and didn't go to welcome at last. In the 23rd—24th years of Jiajing, the country was suffering great drought. Li Shu was sincerely begging for rain by wearing coarse cloth gown and walking under hot sun. He also sold the collected millet in a fair price and opened the offical barn to feed famine refugees, thus saving the lives of many refugees. After his death, his memorial tablet was placed in famous-officials ancestral hall.

民国厚德荣归铭文铜镜
Copper mirror of great-virtue and glory-return inscription in the Republic of China

红娘子投闯王　Miss Hong went and sought refuge with Li Zhicheng

明末饥荒，李信为饥民向县令请求暂时停止征税，却被县令下狱。红娘子知道了，将李信从狱中救出。两人逃出后投奔李自成。

In face of famine at the end of the Ming Dynasty, Li Xin pleaded the magistrate for a temporary shutoff in levying taxes, however he was thrown into prison. Miss Hong saved Li Xin out of cell after knowing this. Successfully escaping, the two went and sought refuge with Li Zicheng.

现代粉彩红娘子投闯王瓷挂盘 Modern famille rose porcelain hanging dish of Miss Hong going and seeking refuge with Chuangwang（Li Zicheng）

五、礼，履也　Propriety（Li）is fulfillment

张端木移风易俗，教育感化百姓［清］　Zhang Duanmu transformed outmoded habit and custom and helped people to change by education and persuasion［the Qing Dynasty］

张端木，字昆乔，上海人，清乾隆七年进士。乾隆十九年上任金华县，诚心为政，宽大与严厉相结合，尤其以教育感化百姓为当务之急。金华邑西有个名为虹路庄的地方，众人皆谓之"荒村"，百姓愚昧，不知礼学。张端木勉励村民读书，并且送文稿给他们，之后这里相继出的文士多于其他村。端木断事如神，违法乱纪的人也不敢使用其伎俩。后离开金华，迁至诸暨。着破衣、乘破车，历五任如一日也。

Zhang Duanmu, with the style name as Kunqiao, was born in Shanghai and achieved the rank of the Scholar in the 7th year of Qianlong in the Qing Dynasty. In the 19th year, he went to Jinhua county for work sincerely. His duty was combined with generousness and strictness, with focus on helping people to change by education and persuasion. There was a place, Honglu village, infamous as a deserted village. Its people were ignorant and had no ideas of rituals. Zhang Duanmu encouraged villagers to read books and sent manuscripts to them, making there more scholars than other villages. Duanmu decided a matter with an excellent judgement, so people who wanted to violate the law and discipline didn't dare to do other tricks. Later on, he moved to Zhuji after leaving Jinhua. He always wore ragged clothes and rode busted car every day during his five terms of office.

六、智，烛也 Wisdom(Zhi) is candle

宋濂勤奋求学 ［明］ Song Lian undertook a diligent study ［the Ming Dynasty］

宋濂，字景濂，号潜溪，祖籍金华潜溪，至宋濂时迁居浦江。宋濂聪敏好学，有"神童"之称，但家境贫寒，买不起书。宋濂小时候常常奔波于邻里邻外，忙着向周围的人借书。小宋濂对书的喜爱已经到了如痴如醉的地步，但是看一遍不过瘾怎么办，就把每一本书都抄写下来。宋濂深知自己总是借别人的书很尴尬，于是他每借一本书都定下一个还书时间，让书主人放心将书借给他。有一次，宋濂又到一个富贵人家里借书看，却吃了闭门羹，宋濂着急了，因为这是一本难得的好书，虽深知十日之内绝对读不完，但还是约定十日之后必定还书，书主人才勉强答应。当时是寒冬时节，到了约定的第十天早晨，漫天飞雪，气温骤降，没有人想要在大街上逗留，更不用说长途跋涉还书，书主人想着宋濂不会履约。当宋濂顶着大雪出现在主人面前时，书主人非常感动，并答应以后宋濂随时可以来借书，而且不用限定还书的日期。因为宋濂勤奋刻苦加上遵守约定，在家境不好的情况下，年幼的宋濂依旧饱读诗书。

Song Lian had his style name as Jinglian and his pseudonym as Qianxi. His ancestral home located at Qianxi, Jinhua and moved to Pujiang after Song Lian was born. As a smart and hard-working kid, Song Lian was reputed as "whiz kid", but he could not afford books due to his poverty. Hence, when he was young, he often went to neighbors to borrow books. But it was not enough for him to read a book for one time. His love for books was so intoxicated that he transcribed every book. He also knew that it was embarrassing to borrow other's books, hence he made a date of returning books with the owner every time. That made the owner feel comfortable and relieved to lend him books. Once, Song Lian went to a rich person's home to borrow books but was refused. Getting worried, Song Lian knew that this book was a god-given good book. So Song Lian promised to return the book in ten days, making the owner reluctantly agree to lend. However, Song Lian was well aware that it was impossible to complete within ten days. When it came to the morning of the 10th day, it was snowy and freezing due to severe winter. No one wanted to hang out in the street, let alone walking long distance to return a book, thought by the rich. But, when the rich opened the gate, he found Song Lian with his head covered with white snow. The rich was so moved that he promised to lend books to Song Lian any time and there was no date of returning any more. Thanks to his industrious style and compliance with his promises, young Song Lian was well read though living in a poor family.

七、信，诚也 Honesty(Xin) is credit

纸币、汇票首先在婺州出现 Paper money and check emerged in Wuzhou

金华素为兵家必争之地。北宋末年，新安江流域爆发方腊起义，杭州与金华之间水路不通，运输不安全，造成了朝廷在金华的驻军缺少军费现钱。

Jinhua was always a place of great military importance. At the end of the Northern Song Dynasty, Fang La uprising burst out in Xinanjiang river basin. That blocked up the waterway between Hangzhou and Jinhua and made the transportation unsafe, resulting in the lack of military expenditure cash for governmental garrison troops in Jinhua.

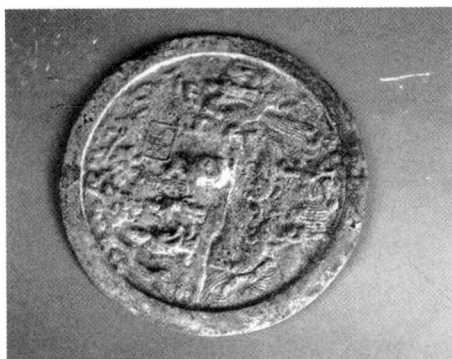

宋金柳毅传书纹铜镜 Copper mirror of Liu Yi passing a message in the Song and Jin periods

宋高宗绍兴元年，政府号召金华的商人把现钱交给婺州官府，官府给予存钱的纸质凭证——关子。金华商人到杭州、绍兴做生意时，可以用这些纸质关子到官府兑取现钱，于是在金华就诞生了中国最早的纸币、汇票之一——关子。

At the first year of Shaoxing, Emperor Gaozong of Song, the merchants of Jinhua was called by the government to hand cash to Wuzhou government. The government would give them a paper voucher showing the deposit, Guanzi. When Jinhua merchants went to Hangzhou and Shaoxing for businesses, they could cash in the government by the paper Guanzi. Hence, one of the earliest paper money and check, Guanzi emerged in Jinhua.

显而易见，纸币、汇票能首先在金华出现，缘于婺州这片富有社会诚信的土壤，基于人与人、婺商与政府之间的信任。

Obviously, the emergence of paper money and check in Jinhua originated from the trust between people and people, between Wuzhou merchants and the government, such trust was deeply rooted in Wuzhou, a land full of social reliance.

民国朱文"实事求是"骨方章 Rectangle bone seal of characters "实事求是 (seeking truth from facts)" in relief in the Republic of China

民国朱文"守口如瓶"青田石随形章 Qingtian naturally-shaped stone seal of characters "守口如瓶 (as close as wax)" in relief in the Republic of China

八、廉，清也 Integrity(Lian) is incorruptness

陈天瑞志行廉洁［南宋］ **Chen Tianrui was clear-handed in his ambition and conducts［the Southern Song Dynasty］**

陈天瑞，字景祥，临海人，咸淳年间任金华知县，有贤能的名声。师从王柏，对理学颇有研究，志行廉洁。宋亡以后，归隐山林，诗文古澹，效仿陶渊明，著有《甲子集》五十卷。

Chen Tianrui, styled himself as Jingxiang, was born in Linhai. During Xianchun period, he shouldered Jinhua county, well-known for his virtuous character. Learning from Wang Bai, he had a good command of Neo-Confucianism and was clear-handed in his ambition and conducts. After the demise of the Song Dynasty, he returned to the remote for ancient and simple life along with poems. He simulated the style of Tao Yuanming and rolled out 50 volumes of *Jiazi*.

辽人物故事纹葵形铜镜 **Sunflower copper mirror of recording people's stories of the Liao Dynasty**

宋朝名臣赵抃，为官清廉爱民　Zhao Bian, famous official in the Song Dynasty, was clear-handed and loved his people(the Song Dynasty)

赵抃因反对王安石青苗法被贬入蜀，孤身一人，仅一琴一鹤相随。

Zhao Bian was exiled to Sichuan due to he went against the Qingmiao Reform of Wang Anshi. He was alone, only accompanied by a koto and a crane.

民国兽钮白文"琴鹤遗风"青田石方章 **Qingtian rectangle stone seal of "琴鹤遗风(koto and crane legacy)" in the Republic of China**

第二部分　传颂·家规
Eulogizing · Family Rules

家风好，则族风好、民风好、国风好。

没有好的家规家风，既难以清白做人，也无法专心做事。

——习近平总书记

Good family tradition will generate outstanding national ethos, folk custom and national customs.

Without good family rules and traditions, it is hard to be an innocent man and concentrate on work.

——General Secretary Xi Jinping

人必有家，家必有训。家规家训，行之于文，烙刻于心，世代相传。"一室之不治，何以天下家国为？"家庭教育、家风文化，关系几代人的命运，是一个家庭、一个家族、一个国家兴衰的关键。我国自古以来就十分重视家庭教育，认为"教，先从家始""正家而后天下定矣"。自古至今，那些为国家、为民族做出贡献的英贤和优秀人物，无不得益于良好的家庭教育。

A man must have a family, and a family must have its rules. Family rules and instructions, written on the paper and engraved in the mind, passed down to generations. "If you cannot govern a family, how can you rule the country?" Family education and family tradition culture, closely linking to the destinies of several generations, serves as the key that decides the ups and downs of a family, a clan and a country. Since ancient times, our country has attached much attention to family education, thinking, "education begins in a family" "when a family tradition goes normalized, the world is peaceful." So far, those excellent men who made contribution to the country and the nation were all raised up by good family education.

家规家训作为我国古代家庭教育中的价值追求和行为准则，是中华儿女一脉相承的主流价值观和处世哲学，成为中华民族优秀文化遗产的重要组成部分。金华家规家训所强调的忠孝、廉洁、礼仪等，都充分体现了中华民族的传统美德，更是我们今天应该继承和发扬的文化遗产，成为当代家庭教育中重振家规家训的力量源泉。

Family rules and instructions, as the value pursuit and code of conduct in ancient Chinese family education, is the mainstream value and philosophy of life inherited by Chinese people, and has become a critical part of Chinese excellent cultural heritage. The loyalty and filial piety, integrity and rituals stressed in Jinhua family rules and instructions fully demonstrate the traditional virtues of the Chinese nation. These cultural legacies should be inherited and developed by people today and become the source of power for revitalizing family rules in modern family education.

一、东阳　Dongyang

1. 蔡氏　The Cais

蔡氏家训　Family Instructions of the Cais

东阳蔡氏源远流长，据史料记载，最早到东阳的是会稽（今绍兴）人蔡喜夫，他在任东晋光禄大夫时，于晋穆帝永和五年（349），为避石赵之乱，在东阳画溪南陇居住。几千年来，蔡氏先辈开疆拓土，不断分别从福建、河南及本省的温州、新昌、诸暨等地到东阳择地休养生息。经统计，蔡氏宗亲如今已在东阳市内的蔡宅、白溪、白水口、南马镇的苞竹园等几十个村居住繁衍。截至 2010 年 11 月 1 日止，有 10535 人，占全市常住人口的 1.31%，为东阳第二十七大姓。

The Cais in Dongyang has a long history. According to historical records, the first man who arrived at Dongyang was Cai Xifu, a Kuaiji person(now Shaoxing). As the Guanglu official of the Eastern Jin Dynasty, he moved to Nanlong, Huaxi, Dongyang in order to hide from the chaos of the later Zhao regime at the 5th year of Yonghe, Emperor Mu of Jin(349). For thousands of years, the ancestors of the

Cais exploited the boundary and chose a place of Dongyang to recover from Fujian, Henan and Wenzhou, Xinchang, Zhuji of the province. After statistical analysis, the ancestors of the Cais have lived and multiplied in scores of villages, Caizhai, Baixi, Baishuikou in Dongyang and Baozhuyuan of Nanma Town. Until November 1st, 2010, the population of the Cais reached 10,535, making 1.31% of Dongyang permanent resident population. Thus, Cai became the 27th popular surname in Dongyang.

蔡氏家规　Family Instructions of the Cais

强修身。君子立身，学于先贤，志在道德，利不可诱，势不能屈，一身正气，一诺千金；言必信，行必果；心胸开阔，慷慨磊落，淡泊名利荣辱；于国以忠，仁爱当先；接人以和，谦恭礼让。立品当以树正气、贬污浊、宏道德、炳情操为要，谨言慎行，不执己见，不谋幺利。族人各宜善事之。

——《蔡氏新订族规家训》

You need to strengthen your body and build good character. A gentleman has to learn from the wise and aims to build good morality. You cannot be lured by interests, or yield to the powerful. You must keep upright and keep your promise. You do what you say and make it best. You need to be open-minded, generous, upright and have no interest in personal reputation and glories. You need to be loyal to your country, and treat others with benevolence. You need to be kind, modest, and humble to others. You have to focus on building uprightness, rejecting muddiness, developing morals and showing good sentiment. You must be careful about what you say and do. Don't be stubborn or seek for private interest. All people in our family should try your best to obey these instructions.

——*Newly Revised Family Rules and Instructions of the Cais*

释义：加强修身养性。君子立身，应该向先贤学习，志在磨砺良好的德行。不受利益诱惑，不因权势欺压而屈服。一身正气，一诺千金。说到做到，要做就做到最好。心胸开阔，慷慨磊落，淡泊名利荣辱。对国家要忠心，对人要仁爱为先。待人和气，谦恭礼让。以树立正气、拒斥污浊、弘扬道德、彰显情操为重，谨言慎行。不顽固偏执己见，不谋求私利。各位族人应当尽量遵守。

《蔡氏新订族规家训》木雕文字版 Woodcut of *the Newly Revised Family Rules and Instructions of the Cais*

清廉布政司蔡安贵　Clear-handed chief secretary Cai Angui

蔡安贵（1386—1451），今东阳市湖溪镇白水口村人，以书法名冠天下，是历史上白水口泉溪蔡氏先人的优秀代表。官居明朝河南布政司，在任以清廉著称，是当时为人称道的清官和著名的文学家、书法家。

Cai Angui（1386—1451）was born in Baishuikou village, Huxi town, Dongyang. Well-known for his calligraphy, he was an excellent representative of the ancestor Cais of Quanxi, Baishuikou in the history. He was the chief secretary of Henan in the Ming Dynasty, famous for his in-corruptness. He also was a reputed honest and upright official and famous litterateur and calligrapher.

蔡安贵自幼天赋聪灵，年少好学，尤其醉心于对文学、书法的研究与刻苦磨砺，终于练得一手好字，写得一手华章。明永乐元年（1403），明成祖朱棣下令修撰汉字大词典《文献大成》以全面完善汉字、汉词。明永乐三年（1405）时，声名远播的书法才俊蔡安贵被明朝内阁首辅、著名学者解缙发现，应试录用招入文渊阁，参加纂修《文献大成》，即后来的《永乐大典》。这是我国古代编纂的一部大型工具书，是中华民族珍贵的文化遗产，是中国古代最大的百科全书。《永乐大典》收录古代重要典籍七八千种之多，上至先秦，下达明初。收录的内容包括经、史、子、集、释庄、道经、戏剧、平话、工技、家艺、医卜、文学等，无所不包。历时 4 年的工作，给予了蔡安贵学习、历练、表现与发展才能的机会。他的杰出成就得到朝廷的认可与赏识：在金殿面圣时，皇帝赐予白银二百锭，补太学生以示嘉奖。明宣德四年（1429）被任命为刑部交趾清吏司主事。嗣后，先后任湖广清吏司主事、四川清吏司主事署员外郎、山东省右参政兵政。正统二年（1437）皇帝亲授河南布政司。

蔡安贵谨记蔡氏祖训，为官一地，清廉一方。无论是在朝廷为官，还是任不同地方的不同职官，都能在当时充满污泥浊水的官场里，挺胸抬头，公正办事，清廉做官，深得民心，也屡受朝廷表彰。

As a gifted person, Cai Angui was studious when he was young, especially enchanted with the learning and training of literature and calligraphy. That made him able to write a good handwriting and excellent articles. At the 1st year of the reign of Yongle in the Ming Dynasty(1403), Emperor Chengzu, Zhu Di ordered to compile the dictionary of Chinese characters, *Wen Xian Da Cheng*, in order to fully perfect Chinese words and phrases. At the 3rd year of the reign of Ming Emperor Yongle(1405), prestigious young talented calligrapher, Cai Angui was discovered by Xie Jin, a famous scholar and cabinet Prime Minister of the Ming Dynasty. Cai Angui was enrolled to the Imperial Library to join the compilation of *Wen Xian Da Cheng*, namely later *Yongle Canon*. This is a large scale reference book compiled in the ancient times, precious cultural heritage of the Chinese nation, and the biggest ancient Chinese encyclopedia. *Yongle Canon* collects seven-to-eight thousand of important ancient classics, ranging from the pre-Qin period to the early Ming Dynasty. Its contents consist of almost everything, like classics, history, philosophy, literature, explanation of Zhuangzi, Taoist scriptures, drama, popular stories, handicrafts, home art, medicine and divination and literature. Four-years working gave Cai Angui the chance of learning, toughening, behaving and developing his skills. His excellent achievements won the recognition and appreciation of the government. When presenting himself before the Emperor in the golden court, he was rewarded 200 ingots of silver, which was used to praise the students of imperial college. At the 4th year of the reign of Ming Emperor Xuande (1429), he was appointed as the chief of appointment and transfer department of Ministry of Justice in Jiaozhi. Hereafter, he also shouldered the chief of appointment and transfer department of Hubei and Hunan, chief assistant director of Sichuan appointment and transfer department, right administrator in army of Shandong. At the 2nd year of the reign of Emperor Zhengtong(1437), the Emperor personally appointed him as Henan chief secretary. Cai Angui always kept the Cais' instruction in mind, so every place he governed had no corruption. Whether he worked in the imperial court, or shouldered different positions in different places, he could stand upright, work fairly and honestly in filthy and corrupted official circles. Hence, he was popular among the locals and often got praised from the court.

蔡安贵热心家乡建设，在祖居地东泉溪首建大宗祠，开辟"泉溪八景"，重建"宝安寺"，建石砌城墙与十字街道，功德无量。是泉溪蔡氏后裔至今念念不忘的先辈楷模。

Cai Angui devoted himself into the construction of his hometown. He firstly built ancestral temple in Dongquanxi, his ancestral homeland. Then, he developed "Quanxi Eight Sights", rebuilt Bao An Temple, renovated stone city walls and crossroads. The benefits people have received from him are beyond measure, thus making him an unforgettable ancestral model for his Quanxi Cai generations.

教育家蔡汝霖 Educator Cai Rulin

蔡汝霖（1868—1916），字雨香，晚号愚公，今东阳市虎鹿镇蔡宅村人。清光绪二十三年（1897）浙江乡试中第四十八名举人，是我国现当代著名的国务活动家、教育家。

Cai Rulin（1868—1916）had his courtesy name as Yuxiang, and he also self-titled himself Yugong in old ages. He was born in Caizhai village, Hulu town, Dongyang. At the 23rd year of the Qing Emperor Guangxu（1897）, he became the 48th successful candidate in Zhejiang provincial examination. He was famous contemporary state-man and educator in China.

蔡汝霖 Cai Rulin

蔡汝霖中举后，面对清晚期的黑暗社会，深感民族之危难，力主变法图新。他满腔热情投入"文事武备"。曾相继任新昌"沃西书院"山长、创立并亲任"浙江武学堂"堂长，努力培养能文能武的建国人才，名噪浙江。公元1903 年，他东渡日本，考察教育，结识了孙中山与光复会元老陶成章，加入了当时致力于振兴国家民族的革命组织"同盟会"。

After passing the exam, he deeply worried about the national calamity and urged for national reform and innovation in face of the dark society of the late Qing Dynasty. With full passion, he devoted himself into "non-military foreign affairs and military talented personnel". He shouldered the principal of Xinchang "Woxi

Academy", and established "Zhejiang Kungfu Academy" and acted as its master. He was committed to cultivating talents proficient in both brainy and brawny activities for the development of the country, making him popular in Zhejiang. In 1903, he went to Japan for an education observation, during which he got to know Sun Yat-sen and Tao Chengzhang, the founding member of Restoration Society. He also joined "Chinese Revolutionary League", a revolutionary organization that aimed to rejuvenate the country and the nation at that moment.

回国后，被聘为金华府中学堂监督（即校长）。同年 6 月，创办《萃新报》。又应好友之邀回东阳家乡筹备扩建县立小学。1905 年，创办"永宁自治高初两等小学堂"。第二年，又与友人在杭州创办金、衢、严、处四府旅杭公学，自任监督。由于成绩卓著，被公推为全浙江教育会干事。1907 年，他返回故里，出任东阳教育会会长。宣统元年开始，任浙江省立七中校长。1911 年，浙江军政府成立，被委任为宣慰使。1913 年，推选为中华民国国会众议院议员。1915 年 4 月出任浙江护国军政治顾问。1916 年阴历十二月十七日病故于蔡宅"听春雨楼"。

After returning China, he was hired as the president of Jinhua Prefecture Middle School. During June of the same year, he established *Cuixin News*. Invited by his friends, he went to Dongyang to prepare the extension of County Primary School. In 1905, he set up "Yongning Autonomous Senior and Junior Primary School". In next year, he built four public schools of Jin, Qu, Yan, Chu in Hangzhou, and he acted as their supervisor. As for his excellent educational achievements, Cai was promoted to be an associate of Zhejiang Educational Association. In 1907, he returned his hometown and acted the president of Dongyang Educational Association. At the 1st year of the reign of Emperor Xuantong, he served as the president of No.7 Zhejiang Provincial High School. In 1911, Zhejiang Military Government was set up, and he was appointed as Xuanwei ambassador. In 1913, he was elected as a senator of House of Representatives of the Congress of the Republic of China. In April, 1915, he played the role of political consultant of Zhejiang National Protection Army. On the 17th day of the 12th lunar month 1916, he died of an illness in "Listen to Spring Rain Building" at home.

蔡汝霖为推翻清朝封建王朝、建立新的国民政治体制，代表人民参政、议政以及建立新的教育机构、教育体制，为教育事业做出了杰出贡献，在当时的浙江乃至全国具有重大的影响力。他是追随孙中山的东阳蔡氏后代，直接亲自参与辛亥革命的早期革命家、国务活动家和著名的教育家。他追求进步、为国家的强盛奔走呼号、扎实工作的精神将永远激励后人，是东阳鹿峰蔡氏后裔学习的楷模。

In order to overturn feudal imperial Qing Dynasty, establish new national political system, represent people to take part in and discuss politics, and set up new educational organizations and systems, Cai Rulin made great contribution to educational sector. He had a huge influence on Zhejiang and even the whole country back then. He was a descendant of Dongyang Cais and a Dongyang person following Sun Yat-sen, an early reformer directly participating in the Revolution of 1911, a state-man and famous educator. The spirit of pursuing progress, urging for national prosperity and steady working will stimulate later generations forever, which is also a role model for the generations of the Cais of Lufeng, Dongyang.

2. 何氏　The Hes

桓松何氏，姬姓，系周成王弟唐叔虞孙韩安王之裔。其裔有以封地为氏；有以改韩为"何"，而从何氏。韩愈所谓"韩、何同姓者"，此也。后汉何元干，幼时就食江南，及长，家于庐江，后子孙遂以为郡望。其后支有迁宣城者。至何处官婺州，遂自宣城奉父而迁至婺。至何慧，仕至中奉大夫，约于乾道七年迁于东阳一都（今卢宅前址）。其后裔分居林头、南上湖、防军等处。其人物，宋有何逵，嘉定进士，生八子，六人出仕。何梦然，进士，官至参知政事；何梦全，官至尚书；何梦祥，进士，官司农卿；何梦申，治《周礼》名于时。元有何正，伯颜屡征不就，避居乌伤。

The Hes of Hengsong, with their family name as Ji, were the generations of King Han'an, the grandson of Tang Shuyu, the younger brother of Zhou Emperor of Cheng. His generations took the name of his fief as their family name, so "Han" was changed to "He", making the birth of the Hes. Han Yu once said, "The Han and The He shared one name", which indicated the above fact. He Yuang an of the later Han Dynasty, lived in Jiangnan when he was young. When he grew up, he moved to Lujiang, a place regarded as their hometown by his descendants. His generations also moved to Xuancheng. When He had an official position in Wuzhou, he went to Xuancheng and made his father move to Wuzhou. He Hui was appointed as Zhongfeng official and moved to Dongyang at the 7th year of the

Qiandao regime（the former Luzhai）. His descendants separately lived in Lintou, Nanshanghu, Fangjun and other places. He Kui in the Song Dynasty was Jiading Scholar and had eight sons, six of whom became officials. He Mengran, a scholar, had the power to participate and discuss political affairs; He Mengquan had the position of Minister; He Mengxiang, a scholar, acted as an official responsible for grains; He Mengshen, was famous for his studies on *Rites of Zhou*. He Zheng in the Yuan Dynasty, refused to be conscripted by Bo Yan for many times and went to live in Wushang.

（何氏）南上湖东村古戏台（俗称"雨台"）
(The Hes) Ancient drama stage of Dongcun, Nanshanghu (commonly known as " Rain Stage")

（何氏）林头悌青阁（The Hes）
Lintou Tiqing Attic

何氏家训　The Hes' Family Instructions

读书必专静，写字必端庄。

出言必谨审，行步必安详。

事亲必承顺，立身必直方。

接物必和易，遇事必商量。

治家必勤俭，为官必清强。

克恃此十者，德业自芬芳。

——何梦然《十必箴言》

Be concentrated and quiet when reading; be dignified when writing; be careful when speaking; be peaceful and steady when walking; be filial and respectful when treating our family; be upright when standing; be kind and easy-going when dealing with people; consult with others when having troubles; be diligent and thrifty when governing the family; be clear-handed and make achievements when being an official. Your virtue and achievement will be presented if you obey these ten rules.

—He Mengran *Ten Necessities*

释义：读书必须专心且安静；写字必须端庄大气；说话必须谨慎小心；走路必须安详稳重；侍奉亲人必须孝顺恭敬；站立必须挺直大方；待人接物必须随和平易；遇事必须与人商量；治家必须勤俭节约；为官必须清廉有为。恪守这十条原则，德行和成就必定会得以彰显。

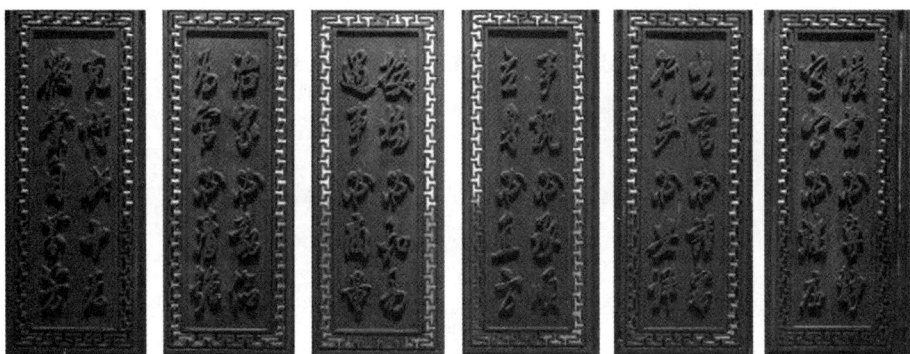

何梦然《十必箴言》木雕文字版 Woodcut *Ten Necessities of He Mengran*

3. 吴宁厉氏　The Lis of Wuning

厉氏家族　The Lis

厉氏，系出炎帝后之烈山氏。姜尚助武王克商，封于齐。传六世厉公即位，子孙遂以所谥之"厉"为氏，以齐国为郡望。又一支，以汉光武时之厉井子厉温，平蛮有功，封"义阳侯"。东阳厉氏之为"义阳"郡者以此。后裔雷甫，避西晋末年之乱，由义阳（河南新野）迁东阳岘山之麓，后为大族。其裔唐宋人才多，代表人物有厉文才，唐贞观进士，任容州都督刺史，为东阳"开五府"之首；厉仲祥，宋武科状元；厉模，官至正卿，累赠少保。

The Lis was the descendants of Emperor Yan, Lieshan clan. Jiang Shang helped Emperor Wu to combat the Shang Dynasty, thus Jiang was given his fiefdom in Qi country. On the enthronement of Lord Li after six generations, Jiang's descendants took Li as their surname and Qi country as their hometown. Another generation came from Li Wen, the son of Li Jing, in the reign of Han Emperor Guangwu. He made great contribution when suppressing barbarians, titled with "Marquis Yiyang". That was the reason why the Lis in Dongyang took "Yiyang" as their hometown. His descendant, Lei Fu moved to Xianshan, Dongyang from Yiyang(Xinye, Henan) to avoid the chaos in the late Western Jin Dynasty, later forming a big family. In the Tang and Song Dynasties, there were many talents in his descendants, like its representatives, Li Wencai, the Scholar during the reign of Emperor Tai Zong, administrator of Rongzhou, the first of the "Five Dongyang Officials" (set up his own mansion and summon advisors). Li Zhongxiang, was the best Scholar in the military test of the Song Dynasty. Li Mu, was supreme chief Minister for administration and finally honored as the teacher of the prince.

瑞霭堂 Rui Ai Hall

吴宁厉氏家训　Family Instructions of the Lis, Wuning

子孙有出仕者，当尽忠补过，不可贪污酷虐，以致亡身及亲。否则，覆饰覆辙，读书何为？其不仕者，忠厚存心，正直是与。或区处公事，当尽心竭力，排难解纷，不可苟且造次，左右其袒，首鼠两端，致误于人，招怨不小。

——《吴宁厉氏家规》

If there were descendants becoming officials, they should be loyal to the country when getting promoted. When being downgraded, they should make up their faults. And, they cannot embezzle, bully people, in order not to harm themselves and get their family into trouble. Otherwise, what can studying bring about if a scholar still does bad things. If not being an official, they should be also upright. When dealing with official business, they should try their best. When mediating disputes, they cannot slight over the matters and cannot show favoritism to any parts or be timid and hesitant, otherwise they will be complained.

—Family Instructions of the Lis, Wuning

释义：子孙中如果有出仕为官的，应当进则效忠国家，退则弥补过失，不可贪污，不可欺凌百姓，以至于害己，甚至连累家人。否则，读书而又犯恶，那么读书又有何用？如果不为官，仍应心存忠厚正直。处理公事时，应当尽心竭力；排解纠纷时，不可苟且敷衍，不可徇私偏袒、畏畏缩缩，否则必会招致怨怼。

《吴宁厉氏家训》木雕文字版 Woodcut *Family Instructions of the Lis*，**Wuning**

最早进士 The first Scholar

厉文才（606—683），字日新，一作安世，号蓉州。横店夏厉墅人。

Li Wencai(606—683), has its courtesy name as Rixin, or Anshi, and was self-titled as Rongzhou. He was born in Xialishu, Hengdian.

厉文才是东阳第一个中进士者，他于唐贞观元年（627）登第，该年全国中进士4人。时李渊立国未久，边地尚未大治，南土夷寇猖獗，尤以荔浦之寇最为嚣张。该地溪峒连绵，行途阻绝，用兵颇难。厉文才临郡征讨，"煦之以皇风，董之以戎略"，期月之间，群盗悉平。

Li Wencai was the first scholar in Dongyong. He passed the imperial examination at the first year of the Zhenguan period of the Tang Dynasty(627), one of four successful candidates at that year. When the Tang Dynasty was just established by Li Yuan, the borderland was in demand of official governing and the barbarians of the south were furious, especially in Lipu. The place was full of streams and caves, which blocked the pathways, making it difficult to conduct military operations. Li Wencai went to combat, "using the imperial education to reclaim these people, and taking military strategies to govern them". Within months, these places were back to peace.

厉文才辞官归里后，捐资鸠工在禹山北麓凿山挖石，兴修水利。沿荆溪（南江）修渠7华里余，引山泉灌溉夏厉墅、湖头陆等地农田一千五百余亩。堰成，百姓甚感厉文才之德，因为他曾都督岭南，于是命名为"都督堰"。

When Li Wencai resigned and returned his hometown, he made a donation to dig mountains for water conservancy projects at the north of Yushan Mountain. Along with Jingxi（Nanjiang），

厉文才 Li Wencai

over 7-li canal was built to irrigate the farmland of more than 1,500 mu, like Xialishu and Hutoulu by diverting mountain water. When the weir was finished, the people expressed heartfelt thanks to his virtue. On account of his former position, Lingnan governor, so the weir was named as "Governor Weir".

二、金华婺城区　Wucheng of Jinhua

盘溪何氏　The Hes of Panxi

金华市婺城区罗店镇后溪河村，古名盘溪村，因发源于北山深处九龙村的盘溪盘绕村庄潺潺流过而得名。自宋代以来，盘溪何氏就在此聚居，至今已历三十余代。不论岁月如何变迁，何氏一族始终勤俭继世、耕读传家，有着"理学世家"的美誉。追古溯今，族中人才辈出，其中南宋理学家何基潜心研学、传业授道，对朱子理学的发展贡献突出，与他的三位学生王柏、金履祥、许谦并称为"金华四先生"，也叫"北山四先生"。在宋、明两代，盘溪何氏共出过9位官员，并有2位官居正二品户部尚书，为官皆务实清廉。近代以来，"何氏三杰"——教育家何炳松、知名爱国人士何德奎、史学家何炳棣更成为

何氏一族长盛不衰、人文基因世代传承的重要代表人物。

Houxihe village, located at Luodian town, Wucheng district, Jinhua city, was named Panxi village in the ancient time, well-known for that the Panxi river, originated in Jiulong village at the deep of north mountain, flowed from village to village. Since the Song Dynasty, the Hes of Panxi gathered in this place and have generated over thirty generations. Despite of time changes, the Hes stick to hardworking and thrifty style and learn how to be a man as well as how to make a living, reputed as "The Family of Neo-Confucianism". Since the ancient times, there were many talents in this family. For example, He Ji, a neo-Confucianist in the Southern Song Dynasty, devoted himself into studying and teaching, making great

何北山祠内牌匾 **Tablets in the North Mountain Hall of the Hes**

contribution to the development of Zhu Xi's neo-confucianism. He and his three students, Wang Bo, Jin Lvxiang and Xu Qian were known as "Jinhua Four Scholars", also "North Mountain Four Scholars". During the Song and Ming Dynasties, there were 9 officials in this family, and two of them were the second class ministers of revenue. Both of them were steady and clear-handed. Since the modern times, "Three Talents of the Hes", educator He Bingsong, famous patriot He Dekui and historian He Bingdi are important representatives standing for the flourishing humanity genes of the Hes.

何北山祠 **North Mountain Hall of the Hes**

何氏家训（摘要）　Family Instructions of the Hes（excerpts）

何基一生"尚礼义不尚权谋"、专求"内圣"的贤德圣行，被子子孙孙树为楷模。后人以何基的思想品德和行为准则为蓝本，编著了本家族的家规——《何氏家训》。

He Ji "pursued for rites and moralities instead of power and conspiracy", only studied the good virtue and wise behaviors of "internal saints", reputed as a model by his descendants. Later generations took the morality and code of conduct of He Ji as the bluepoint to write their family rules: *Family Instructions of the Hes.*

《何氏家训》共十条，涉及祭祀礼仪、父母长辈、子孙教育、冠婚丧祭、生活学习、为人处世等方面，详尽规范了家族成员的修身处事、待人接物之道，而其中又以第六条、第七条最能凸显何氏家族"明德惟馨、贤廉传家"的风范。

Family Instructions of the Hes consisted of ten rules, including sacrificial rites, parents and elders, children education, coming-of-age ceremony, wedding, funeral and sacrifice, living and learning and social communication. It regulated in detail the moral-character cultivation and inter-personnel communication. Article 6 and 7 in particular, highlighted the demeanor of "good virtue likes refreshing fragrance, virtuous and talented people can flourish the family".

第六条　凡入学入仕等事，祠给衣装。告祠之日，合用鼓乐导拜……祭品随分其余，仪节酌宜而行。

When someone enters the school or becomes an official, the hall will provide him expenses for clothing and travelling. On the day he leaves, people will use drums and other instruments to bid a farewell. Sacrificial offerings are based on convention while rituals are conducted after careful consideration.

这是何氏家族对读书和为官者的表彰和重视。养贤才乃全族人的本分，供养读书，不遗余力，但同时强调"礼仪酌宜而行"，告诫家族子弟要注重节俭。

This was the praise and attention of the Hes to reading and officials. It was the duty of the whole family to raise up talents, support children to study by sparing no efforts. "Rituals are conducted after careful consideration" is stressed at the same time to warn family members to focus on frugality.

第七条　凡族中族长房长公长，须选有德有行素为服众者，立之。如所立之长或有假公济私，妄加人过失以报私忿者，罚银入祠……出别议贤能者，立之。

When choosing the leaders of the family, we must choose the virtuous and upright to meet others' need. If the leader abuses the public trust, treat people unjustly to revenge for private grudge, he will be fined by the hall. We will decide to choose other talents to be the leader.

议贤能规定家族要以贤能者为首，选贤士要严格按照贤德标准公推公选，并通过公开评议，杜绝假公济私，做到公私分明。古人以"立德、立功、立言"为三不朽，而在何氏家族的理念中，为人处世的最高成就就是立德。

The discussion of talents regulating the family should give priority to the talented. The election should strictly follow the virtuous standard by the public. Through public discussion, the use of public offices for private gain was forbidden for being scrupulous in separating public from private interests. Ancient people took three imperishables "to set one's virtue, to set one's merit, to expound one's ideas in writing". However, in the philosophy of the Hes, the highest achievement of social communication is to set the virtue.

何基　He Ji

何基（1188—1268），字子恭，号北山，罗店后溪何（河）人，南宋理学家。1192年，年幼的何基随父亲何伯翼到江西任上。当时，何伯翼的顶头上司是朱熹的女婿黄榦。在何伯翼的引导下，何基和哥哥何南拜黄榦为师，开始接受正宗的儒家理学文化教育。

He Ji（1188—1268）, had his courtesy name as Zigong, and self-titled Beishan. He was born in Houxihe, Luodian. As a neo-confucianist, he, together with Wang Bo, Jin Lvxiang and Xu Qian were reputed as "North Mountain Four Scholars". At 1192, young He Ji went to Jiangxi with his father He

何基 **He Ji**

Bowei, who took work there. At the moment, the immediate supervisor of He Bowei was Zhu Xi's son-in-law, Huang Gan. Guided by He Bowei, He Ji and his brother, He Nan took Huang Gan as their teacher, beginning to study formal neo-Confucianism.

1264 年，宋理宗赐封何基为中奉大夫（正四品），金华开国男（爵）等。1265 年，何基推辞掉各种邀请，悉心钻研儒学、教授门生。主要著作有《大学发挥》《中庸发挥》《易系辞发挥》等。现在仅存《何北山遗集》4 卷。

At 1264, the Song Emperor Zhao Yun appointed He Ji as Zhongfeng official (the forth rank) and Jinhua founding baron. At 1265, He Ji rejected all invitation and focused on studying Confucianism and teaching his students. His main works included, *Development of Great Learning*, *Development of Doctrine of the Mean*, and *Development of Great Appendix of Book of Changes*. Only four volumes of *He Beishan Collection* are remained now.

何基死后，门生尊重老师的意愿，用士礼为他下葬，而不是官葬。1275 年，宋恭帝赵㬎特颁圣旨赐何基谥号文定公。雍正二年（1724），作为大儒的何基，画像被请入各地孔庙供奉。

After his death, his students buried him following the rituals of burying a scholar, rather than an official. At 1275, Song Emperor Gong, Zhao Xi specially awarded He Ji with posthumous title, Wen Ding Gong, through imperial edict. At the second year of the reign of Emperor Yongzheng(1725), as a well-educated scholar, He Ji's portraits were enshrined and worshiped in Confucian temples across the country.

三、兰溪 Lanxi

1. 八卦村诸葛氏 The Zhuges of Bagua village

兰溪诸葛村，位于浙江省金华市兰溪市西部，是全国诸葛亮后裔最大聚居地。现有保存完好的明、清古建筑 200 多处，并按九宫八卦设计布局，是中国古村落、古民居的典范，也是浙江古文化的三大标志之一，被费孝通先生誉为"八卦奇村，华夏一绝"。

Zhuge village of Lanxi is located in the west of Lanxi, Jinhua, Zhejiang, the biggest habitation for the descendants of Zhuge Liang in the country. There are over 200 intact ancient buildings of the Ming and Qing Dynasties. They are designed according to nine directions and eight-trigrams, making the village an example for ancient Chinese villages and folk houses. It is also one of three great symbols of

Zhejiang ancient culture, reputed as "eight-trigrams magic village is a unique place in China."

元代中期，诸葛亮第 27 世孙迁居至此，诸葛遗风从此生生不息。一千多年来，诸葛亮伟大的人格力量和崇高的精神品质，尤其是他留传后世的《诫子书》，教育和激励着一代又一代后世子孙。明清两代，诸葛村就出进士、举人、各类贡生 59 人，正七品以上官员 23 人。一代又一代诸葛裔孙秉承先祖遗风，为官者公而明察、廉而生威、严于自律、不受不污，为民者笃守品行、勤劳务实、积极进取、自强不息，发展和创造了许多新的业绩，被世人广为称颂。

At the midterm of the Yuan Dynasty, the grandson of the 27th descendant of Zhuge Liang moved to here, thus making the remained lifestyle of Zhuge live and grow in nature. Over one thousand years, the great power of personality and noble spirit of Zhuge Liang, especially his *An Admonition to My Son*, educated and encouraged from generation to generation. During the Ming and Qing Dynasties, there were 59 excellent scholars in different fields as well as 23 officials of the seventh rank and higher. Generations of Zhuge people inherited the remained lifestyle of Zhuge Liang. Among them, officials were fair in supervision, clear-handed, strictly self-disciplined, incorrupted, while common people obeyed the moral rules and strove constantly for self-improvement and were hard-working, steady and self-reliant. They developed and created many new achievements, widely eulogized by people.

诸葛八卦村 **Zhuge Bagua village**

诸葛氏家书家规　Family letters and family rules of the Zhuges

作为三国时期著名的政治家、军事家、思想家，诸葛亮位及蜀国军师、丞相，政务繁忙，但他不忘教诲子孙、外甥，《太平御览》《诸葛亮集》中收录其《诫子书》《又诫子书》《诫外甥书》等家书，后人统称为"诸葛亮家书"。

As famous politician, strategist and ideologist of Three Kingdoms period, Zhuge Liang acted as military counsellor and Prime Minister of the Shu Kingdom. Though busily engaged in political affairs, he still taught his descendants and nephews. Family letters like, *An Admonition to My Son*, *Second Admonition to My Son* and *An Admonition to My Nephew*, collected by *Taiping Imperial Encyclopedia* and *Essays of Zhuge Liang* were generally called "family letters of Zhuge Liang" by later generations.

诸葛后裔在长期秉承祖德家风的过程中，又形成了一套完备且十分严格的家规家训，并刊载在《宗谱》卷首，以此树立族人的行为规范和道德准则，被称为《诸葛氏家规》。《诸葛氏家规》最早形成于宋元，完善于明清，共计15条，内容涉及为人处世的方方面面，从修身立德、课书学艺、为人处事到男子冠礼、婚姻丧事、敬宗祭祖、父祖忌辰、家庭伦理、田产维护，乃至言行举止、生活细节等都有严格规定，明确提倡什么，反对什么，禁止什么，并且定有罚则，便于执行。这些家规，对于调节家族内部的伦理关系、贫富关系，凝聚家族，和睦乡里，规范子孙操行，具有相当大的约束力和影响力。

The offspring of Zhuge Liang formed a complete and very strict family rules and family instructions when long inheriting the ancestral morality and family lifestyle. These rules and instructions were printed at the beginning of *Genealogy*, aiming to build behavioral norms and moral standards for family members, known as *Family Rules of the Zhuges*. It was firstly drafted in the Song and Yuan Dynasties and completed in the Ming and Qing Dynasties. It had 15 rules that referred to all aspects of social communication, ranging from self-development, moral composition, learning, social communication, male coming-of-age ceremony, wedding and funeral, ancestor worship, ancestral death-day, family ethic, farmland and property maintenance to behaviors and details of life. It proposed strict rules on the above aspects, which clearly pointed out what to advocate, what to go against and what to forbid. It also formulated punishment for a smooth implementation.

Such rules posed huge constraining force and influences on the regulation of the ethical relation within the family, relationship between the rich and the poor, cohesion of the clan, harmony with the neighborhoods as well as constraints of descendants' behaviors.

家规——《诫子书》 Family Instructions—*An Admonition to My Son*

夫君子之行，静以修身，俭以养德。非淡泊无以明志，非宁静无以致远。夫学须静也，才须学也，非学无以广才，非志无以成学。淫慢则不能励精，险躁则不能治性。年与时驰，意与日去，遂成枯落，多不接世，悲守穷庐，将复何及！

The behaviors of a virtuous talent can only be achieved through disposition cultivation by meditation and morality development and by being respectful to others and thrifty to self. Your ambition can not be showed if you cannot be serene in mind with few desires, your great dreams cannot be achieved if you cannot be peaceful. Learning requires being serene in mind, while a great man must learn how to study. Without learning, you cannot improve yourself, without goals, you cannot achieve anything. You will grow older as time flies, and your ambition will be gradually worn down. Finally your energy will die out and you learn nothing. And what you have learned is not useful in the society, till then you will sadly sit in your unpromising home. It is too late to regret then.

释义：一个德才兼备者的操行，以静心来修炼性情，以恭俭来涵养品德。做不到恬淡寡欲就不能表明志向，做不到平和宁静就不能实现远大理想。学习需要心静，成材必须学习。不学习就无法增长才干，没有志向就不可能学有所成。年龄会同时光一起逝去，意志会随岁月一天天消磨，最后精力衰竭而学识无成，大多不被社会所用，到那时悲哀地守着没有前途的家，即使后悔也来不及了。

诸葛亮 Zhuge Liang

诸葛亮（181—234），字孔明，号卧龙（也作伏龙），汉族，徐州琅琊阳都（今山东临沂市沂南县）人，在南阳卧龙岗隐居长达十年。三国时期蜀汉丞相，杰出的政治家、军事家、散文家、书法家、发明家。在世时被封为武乡侯，死后追谥忠武侯，东晋政权因其军事才能特追封他为武兴王。其散文代表作有《出师表》《诫子书》等。曾发明木牛流马、孔明灯等，并改造连弩，叫作诸葛连弩，可一弩十矢俱发。于建兴十二年（234）在五丈原（今宝鸡岐山境内）逝世。刘禅追谥其为忠武侯，故后世常以武侯、诸葛武侯尊称诸葛亮。诸葛亮一生"鞠躬尽瘁，死而后已"，是中国传统文化中忠臣与智者的代表人物。

Zhuge Liang(181—234) styled himself as Kongming and self-titled himself as Wolong(also Fulong). As a Han person, he was born in Yangdu, Langya, Xuzhou(Yi'nan county, Linyi, Shandong). He had lived a seclusive life in Wolonggang, Nanyang for ten years. He was the Prime Minister of the Shu and Han States of Three Kingdom period, excellent politician, military strategist, proser, calligrapher and inventor. When he was alive, he was appointed as Wuxiang marquis, and after his death, he was conferred as Zhongwu marquis. And he was conferred as King Wuxing due to his military excellence during the regime of the Eastern Jin Dynasty. His famous prose were, *Memorial on Sending Out the Troops* and *An Admonition to My Son*. He also invented wooden oxen and mobile horses and Kongming lantern, and reconstructed repeating crossbow, also named Zhuge repeating crossbow, which can shoot ten rows once. He died at Wuzhangyuan(within Qishan, Baoji) in the 12th year of the reign of Emperor Liu Chan(234). Liu Chan conferred him as Zhongwu marquis, hence later generations called Zhuge Liang as Wu marquis or Zhuge Wu marquis. Zhuge Liang spared "no effort in the performance of his duty and devoted all his life to his country and people", who was a representative of loyal official and wise man in the traditional Chinese culture.

2. 龙门范氏　The Fans of Longmen

兰溪范氏是宋代名丞范仲淹的后裔之一。范仲淹之孙范正路自北宋末由苏州迁入兰溪龙门，为迁兰溪"龙门范氏"始祖。南宋时分迁至厚仁、里范、芷芳岗、彩衣堂、里仁范洞源等村落，明时分迁到范宅。现已繁衍几十代子孙，其间名人辈出。南宋一朝，共出进士5名，举人19人。宋、元、明、清四朝，龙门范氏为官入流者共59人，其中三品以上高官三人，"无地起楼台"的宋代丞相范锺，大理寺少卿范镕，殿中侍御史范处义，香溪"一门双柱国，十子九登科"、彩衣堂孝子范宠，范宅村现代"浙江省十佳孝子"范庆如，这

些都是古今典范，成为人们心目中的楷模。在兰溪黄店范宅、香溪保卫村、灵
洞洞源村都建有范氏祠堂。

The Fans of Lanxi was one branch
of the generations of Fan Zhongyan,
famous Prime Minister of the Song
Dynasty. The grandson of Fan
Zhongyan, Fan Zhenglu moved to
Longmen, Lanxi from Suzhou at the late
Northern Song Dynasty, which was the
ancestor of "the Fans of Longmen" of
Lanxi. His offspring later moved to

范钟铜印 **Bronze Seal of Fan Zhong**

villages like Houren, Lifan, Zhifanggang, Caiyitang, Fandongyuan of Liren in the
Southern Song Dynasty, and to Fanzhai in the Ming Dynasty. Now this family had
dozens of generations, among which there were many famous people. Just in the
Southern Song Dynasty, this family had 5 imperial scholars and 19 first-degree
scholars. There were 59 officials in the Song, Yuan, Ming and Qing Dynasties.
Among them, there were three senior third-rank officials, like "no place for high
buildings" Fan Zhong, Prime Minister of the Song Dynasty; Fan Rong, official of
Dali Temple; Fan Chuyi, imperial censor; Xiangxi "two prime ministers in a
family and nine scholars in ten"; Fan Chong, filial son of Caiyitang; modern
"Zhejiang ten best son" Fan Qingru. These people are good examples from the
ancient to the modern and paradigm in the public mind. The Fans' ancestral halls
are built at Fanzhai of Huangdian, Lanxi; Baowei village of Xiangxi; and
Dongyuan village of Lingdong.

《瀫西龙门范氏宗谱》之《家规》（摘选） *Family Rules of The Fans'*
Genealogy of Longmen, Huxi （excerpts）

孝父母。夫父母乃生身之本，如天如地。

Be filial to parents. Parents are the basis of my birth. The loving-kindness of
parents is as big as heaven and land.

释义：孝敬父母。父母是我身体出生的根本，父母之恩如天如地。

友兄弟。夫兄弟者，父母一体而分者也，如手足然。兄勿以长凌幼年，弟

范宅永思堂 Yongsi Hall of Fanzhai

勿以卑犯尊。

Be kind to brothers. Brothers are born by the same parents, just like the hands and feet. Elder brother should not bully the younger one, while the younger should not offend the elder.

释义：友爱兄弟。兄弟是由父母一体而生的，亲如手足。哥哥不能以年长欺负弟弟，弟弟也不能以小犯大。

别夫妇。夫夫妇乃人伦之本，风化之原，家之盛衰由之而致。必当相敬如宾，相爱如友。

Regulate couples. Couples are the foundation of ethical rules and the source of social customs education. They also have influences on the ups and downs of a family. They should respect each other just as they do to guests; care each other just as they do to friends.

释义：规范夫妻。夫妻是人伦规矩的根本，风气教化的源头，家庭兴衰受他们所影响。一定要相互敬爱如宾客一样，相互关爱如朋友一样。

信朋友。夫朋友为五伦之一，自古所重。必择直谅多闻、品行端方者，与之交焉。

Trust friends. Friendship is one of five human relations, which has been important since ancient times. Make friends with honest, upright and well-informed people.

释义：信任朋友。朋友是五条人伦之一，自古以来很重要。一定要选择正直诚信、见识广博、品行端正的人，与他交朋友。

教子孙。夫子孙乃承先启后之人，所关非小。必教之诚之，以成其名。若不正之於始，则为习俗所染，而德行日秽矣。

Educate offspring. Offspring are important heirs of ancestral wisdom. He must be taught and warned to grow and become famous. If be bad-educated at first, he would be influenced by unhealthy social customs, thus worsening his morality and behaviors day after day.

释义：教育子孙。子孙是承先启后的传人，关系重大。一定要对他进行教育教诲，使他成长成名。如果在开始就教育不好，就会被不好的风气所污染，品德、操行就会一天比一天差了。

范锺　Fan zhong

范锺（1171—1248），字仲和，里范村人。南宋嘉定二年（1209）进士。历官武学博士，知徽州。召赴朝廷，迁尚书右郎官兼崇政殿说书。一次进对，理宗说："仁宗时甚多事。"锺对道："仁宗始虽多事，乃以忧勤致治。徽宗虽无事，余患至于今日。"理宗悦。不久迁吏部郎中兼说书，又迁秘书少监、国子司业兼国史编修，拜起居郎兼祭酒，迁兵部侍郎兼给事中，权兵部尚书兼侍读。嘉熙三年（1239）拜端明殿学士，签书枢密院事。次年授参知政事。淳祐四年（1244）知枢密院事，翌年特拜左丞相兼枢密使，封东阳郡公。时朝纲不正，危机四伏。锺呈《时政十疏》，提出"正君心，定国本，别人才，谨王言，节邦用，计军实，敦士习，清仕涂，结人心，应天命"十条兴国大计，深得理宗嘉纳。多次乞归以保晚节。淳祐六年（1246）辞官。锺为官多年，位居一品，因病回归，竟无钱于故里置地造宅，只得住金华驿馆。故史称："无地起楼台丞相"。淳祐八年（1248）卒于驿馆，赠少师，谥文肃，葬白露山南麓。《宋史》评："锺为相，直清守法，重惜名器，虽无赫赫可称而清德雅量。"著有《礼记解》等。

Fan Zhong(1171—1248) had his courtesy name as Zhonghe and was born in Lifan village. At the second year of Jiading reign, he became a successful candidate of imperial examination. He acted as Doctor in martial arts, governing Huizhou. Called to work in the court, he was promoted to be a high-class assistant Minister and storyteller of Chongzheng Hall. During one expostulation, Emperor Lizong of Song said, "Emperor Renzong of Song was busy in dealing with businesses." Zhong replied: "although Emperor Renzong was busy, he still concerned about his country. However, Emperor Huizong was idle but had so many worries till now." The Emperor was delighted. Soon, he was promoted to be a Minister of War and Minister of supervision, then supervisor of books, high-class teacher and compiler of national history, then Qijulang and scholar, and high official of War and a teacher of Emperors. At the third year of Jiaxi reign(1239), he became a scholar of Duanming Hall and officer in Mishuyuan. Next year, he was ratified to participate and discuss political affairs. At the fourth year of Chunyou reign (1244), he became high official of Shumiyuan, and was Prime Minister and Shumi ambassador in next year. He was also titled as Lord of Dongyang County. At the moment, the law of an imperial court was unstable and the country was in danger. Fan Zhong presented *Ten Suggestions to Politics* to the Emperor, proposing ten suggestions for national prosperity, "put the Emperor's heart in a right place, solidate national foundation, differentiate talents, Emperor should be careful about what he says, save national expense, calculate captured military reserve, correct scholar ethos, clean official court, unite people across the country and be in line with fate." This was awarded by Emperor Lizong. He had pleaded for returning home to maintain his integrity in his later years. At the sixth year of Chunyou reign(1246), he quit office. Being an official for years, he became a first-class official. When he returned hometown due to illness, he had no money to buy a piece of land for house building and could only live in Jinhua posthouse. So as the history recorded, "a Prime Minister had no land to build a house". At the eighth year of Chunyou reign(1248), he died in the hotel, and was bestowed as Shaoshi and posthumously titled Wensu and buried on the southern slope of Bailu Mountain. *History of the Song Dynasty* recorded, "Fan Zhong, as a Prime Minister, was upright, clear-handed and obeyed laws. He valued talents. He had good morality and great generosity even if he didn't make great achievements." He wrote out *Explanation on the Book of Rites* and other books.

3. 后金村金氏 The Jins of Houjin village

据《金氏宗谱》记载，金氏先祖被封于项地，因此以项为氏，后来项伯归汉，赐刘姓。五代时，避吴越王钱镠之讳，改金姓。十一世祖金履祥，因家住仁山之下，学者称为"仁山先生"。桐山后金金氏、长乐金氏，都是金履祥的后裔。金履祥死后的七百年间，德仁文风浩气长存，科甲不断，才子辈出。位于桐山后金村的仁山书院就是当年金履祥隐居仁山下著书讲学的场所，占地面积为 600 平方米，现存建筑为清代风格，结构宏伟、布局得当、保存完整。2005 年公布为省级文物保护单位。近年来，桐山后金村充分发挥本村孝道、理学文化典型的作用，开设了仁山书院纪念馆，积极开展以青少年"四好"教育为主的公民道德教育，以继承弘扬贤孝美德。如今，尊老敬老成风，邻里纠纷、赡养老人纠纷等从未发生过。

According to *The Jins' Genealogy*, Xiang place was enfeoffed to the ancestors of the Jins, hence they took Xiang as their surname. Later, Xiang Bo went back to the Han regime, bestowed with the surname Liu. During the reign of the Five Dynasties, Xiang Bo changed his surname to Jin in order to evade the taboo of Qian Liu, Emperor Yue of Wu country. The 11th ancestor, Jin Lvxiang was called by the scholars as "Mr. Renshan" for he lived at the foot of Renshan. The Jins of Houjin, Tongshan, and the Jins of Changle were the offspring of Jin Lvxiang. During the seven hundred years after Jin Lvxiang died, virtuous and benevolent style of writing and imperishable noble spirit were inherited by this family, cultivating scholars and gifted youth one after another. Renshan Academy, located at Houjin village, Tongshan, was the place where Jin Lvxiang wrote books and taught students when he lived in seclusion at Renshan. This place had a coverage of 600 square meters. Its existing buildings were of Qing style, with majestic structure, proper layout and complete preservation. It was designated as provincial cultural relic protection site in 2005. Over recent years, Houjin village, Tongshan has fully advocated filial piety of the village and set an example of Confucian culture. It established memorial hall of Renshan Academy and actively conducted public ethical education aiming at "four excellence" education for teenagers, in order to inherit and develop sage and filial virtues. Till now, respecting elders has become a fashion and there is neither neighborhood dispute nor old-age endowment dispute.

仁山书院匾额 Horizontal inscribed board of Renshan Academy

仁山书院内景 Indoor setting of Renshan Academy

金氏家训（八行提要） The Jins' Family Instructions（eight-line summary）

孝善父母	友善兄弟
睦善内亲	姻善外亲
任善朋友	恤善州里
忠知君臣之义	和达义利之分

Be filial to parents	Be kind to brothers
Be good to relatives	Be nice to the relatives
Trust friends	Sympathize the local
Be loyal to the Emperor	Be serene in quest for morals and interests

金履祥　Jin Lvxiang

金履祥（1232—1303），字吉父，号次农，自号桐阳叔子，兰溪（今浙江省兰溪市桐山后金村）人。宋、元之际的学者。为浙东学派、金华学派的中坚，"北山四先生"之一，被学者尊称为仁山先生。

Jin Lvxiang（1232—1303）, had his courtesy name as Jifu and self-titled himself Cinong and Tongyangshuzi. He was born in Lanxi（Houjin village, Tongshan, Lanxi, Zhejiang）and a scholar living at the turn of the Song Dynasty and Yuan Dynasty. He was the backbone of Zhedong school and Jinhua school, one of "North Mountain Four Scholars", and reputed as Mr. Renshan by scholars.

金履祥，先祖原姓刘，因避讳吴越王钱镠同音名，故改姓金。从小好学，初受学于王柏，后又学于何基，造诣益深，凡天文、地形、礼乐、田乘、兵谋、阴阳、律历之书，无不精研。时值南宋末年，政治动荡，虽绝意仕进，但未忘忧国。元兵围攻襄樊，履祥献策朝廷，建议以重兵由海道直趋燕蓟，且备叙海舶所经地形，历历可据以行，然未被采纳。
德祐初年，南宋朝廷以迪功郎、史馆编校等职召
任，坚辞不受。寻应严州知州聘，主讲钓台书院。
宋亡，筑室隐居金华仁山下，讲学著书，以淑后
进，许谦、柳贯皆出其门。元大德七年（1303）
卒，至正年间谥文安。

金履祥 Jin Lvxiang

The ancestor of Jin Lvxiang had his original
surname as Liu, in order to avoid the homophone
Liu in Qian Liu, Emperor Yue of Wu, changed the
surname into Jin. Jin Lvxiang was studious when he
was young, firstly studying from Wang Bo, later He
Ji. His knowledge was endless, including astronomy,
geography, rituals, farmland, military strategy,
yin-yang and laws as well as historical books. During the late Southern Song
Dynasty, in face of turbulent political status, he concerned about his country even if
he never wanted to be an official. When Yuan soldiers besieged Xiangfan, Lvxiang
proposed his solution to the court, suggesting that a great amount of soldiers must
go straight to Yanji through seaway. In his proposal, the landscape where the boats
would go through was described in details, making this plan feasible. However, this
proposal was rejected by the court. At the first year of Deyou reign, the court of the
Southern Song Dynasty wanted to appoint him as Digonglang and compiler of
National Archives, but Jin refused firmly. Invited by Zhizhou official of Yanzhou,
he taught students in Diaotai Academy. After the Song Dynasty ended, he built a

house and lived in seclusion at the
Renshan, Jinhua. He taught and wrote
books to cultivate later young talents. Xu
Qian and Liu Guan were both his
students. He died at the seventh year of
Yuan Dade(1303)and got his posthumous
title Wen'an at Zhizheng reign.

金履祥墓 Jin Lvxiang's tomb

四、磐安 Pan'an

1. 榉溪村孔氏 The Kongs of Juxi village

南儒圣地——磐安孔氏家庙 Southern Confucian Shrine—Pan'an Kongs' Family Temple

磐安孔氏家庙由宋代朝廷为"孔氏婺州南宗"（也称作"婺州南孔"）即孔子第四十八世孙孔端躬后裔赐建，现为国家级重点文物保护单位，位于磐安县磐峰乡榉溪村。榉溪村目前是孔子后裔在南方最大的聚居地之一。

Kongs' Family Temple was bestowed by the royal courts for "Kongs' Wuzhou South Master"(also "Wuzhou South Kong"), Kong Duangong, the 48th generation of Confucius. It located at Juxi village, Panfeng, Pan'an county, one of the biggest southern settlement for Confucius's offspring, now a national key cultural relics protection unit.

现存磐安孔氏家庙坐南朝北，是明末清初时期的建筑，门口匾额上"孔氏家庙"四个字依稀可辨。磐安孔氏家庙最早建于南宋宝祐年间，即1253年到1258年之间。当时宋理宗给予"婺州南孔"五级恩典，其中一级恩典就是在榉溪南岸杏檀园赐造至圣家庙。清初，榉溪一带发生农民起义，朝廷派兵镇压，烧杀无度，造成"十年不闻鸡犬之声"，家庙毁于兵乱战火。后来，家庙由孔氏族人集资重建，至今保存较为完整。孔氏家庙古朴宏伟，堂构考究，整座建筑由门楼、戏台、前堂、穿堂、后堂组成，左右对称，布局严谨，气势恢宏，朴实森严。

The existing Pan'an Kongs' family temple faced north. It was a building of late Ming and early Qing Dynasties. Four words "孔氏家庙(The Kongs' Family Temple)" engraved on the inscribed board at the gate were till just recognizable. This family temple was firstly built at the Southern Song Baoyou reign, namely from 1253 to 1258. At the moment, Song Emperor Lizong gave "Wuzhou South Kong" five graces, one of which was to build holy family temple at Xingtan park of the southern bank of Juxi. At the early Qing Dynasty, peasant uprising broke out in Juxi, so the court sent troupes to suppress and killed many people as well as burned many places. This suppression caused "no chicken and dogs barking within ten years" and destroyed the temple. Later, the family members of the Kongs raised funds to rebuild this temple, which was well-preserved till now. The Kongs' family temple was of primitive simplicity and exquisite inside structure. This building consisted of gate-tower, drama stage, antechamber, enterclose and back hall. It featured bilateral symmetry, rigorous layout, solemn and magnificent style and simplicity as well as heavily-guarded atmosphere.

榉溪村 Juxi village

孔氏祖训箴规 The Confucian Ancestral Instructions

一、春秋祭祀，各随土宜，必丰必洁，必诚必敬，此报本追远之道，子孙所当知者。

During the spring and autumn sacrifice, the sacrificial offerings were local products, abundant and clean. People must be sincere and respectful. This was to repay kindness and commemorate ancestors. Offspring should know about this.

二、谱牒之设，正所联同支而亲一本，务宜父慈、子孝、兄友、弟恭，雍睦一堂，方不愧为圣裔。

As suggested by Genealogy, we were all connected, so father must be kind, sons must be filial, elder brother must be friendly, younger brother must be respectful to others, making a harmonious family. Thus, they can be called the descendants of the holy man.

磐安孔氏家庙 **The Pan'an Kongs' Family Temple**

三、尊儒重道，好礼尚德，孔门素为佩服，为子孙者，勿嗜利忘义，出入衙门，有愧先德。

Respect Confucian knowledge and pay attention to what it expresses. Advocate personal etiquette and morality. The Confucians were a good example. Being its offspring, you cannot be addicted to interests and forget all moral principles when being an official. This will bring shame to ancestral morality.

四、孔氏子孙流寓各府州县，朝廷追念圣裔，优免差徭，其正供国课，只凭族长催征。皇恩深为广大。宜各踊跃轮将，照限完纳，勿误有司奏销之期。

Confucian offspring lived in different provinces and counties. The imperial court regarded them as the descendants of holy man, thus they were exempted from corvee, only required to pay taxes, which were levied by the patriarch. So graceful was the Royal Court that the offspring of Confucius should actively pay the taxes on time, so as not to delay local government's collection report to the court.

五、谱牒家规，正所以别外孔而亲一本，子孙勿得勾相眷换，以混来历宗支。

Family rules in Genealogy help people identify our family members, and the offspring cannot marry other family members to disorder the family.

六、婚丧嫁娶，理论守重，子孙间有不幸再婚再嫁，必慎必戒。

Weddings and funerals should be in line of integrity. If they are to remarry out of misfortune, they should be cautious.

七、子孙出仕者，凡遇民间词讼，所犯自有虚实，务从理断而哀矜勿喜，庶不愧为良吏。

When dealing with civil legal cases, those official offspring should be reasonable and feel compassion for victims, and shouldn't crow over. Thus he can be called good official.

八、圣裔设立族长，给与衣项，愿以总理圣谱，约束族人，务要克己秉公，庶足以族望。

Confucius's descendants set the leader of the family, and give him the crown. He is willing to manage the Confucius's genealogy to constrain family members. He must repress the private for the public and build family reputation.

九、孔氏裔孙，男不得为奴，女不得为婢，凡有职官员不可擅辱，如遇大事，申奏朝廷，小事仍请族长查究。

Among the descendants of Confucius, the men cannot be servants and the women cannot be maidservants. They cannot insult any officials. If dealing with big matters, they should submit proposals to the imperial court, while the small one requires the survey of the clan leader.

十、祖训宗规，朝夕教训子孙，务要读书明理，显亲扬名，勿得入于流俗，甘为下人。

Everyday, the offspring should be educated by ancestral instructions and rules. They must study and be reasonable, making their parents well-known for their fame or power. They must not follow current fashion and be willing to become a common person.

孔端躬　Kong Duangong

孔端躬（？—1138），字子敬。原籍山东曲阜厥里，系孔子四十八世裔孙，少力学，登进士第。宋宣和三年（1121）授承事郎、大理寺评事。持身清白，除暴安良，吏畏其威，人怀其德。建炎四年（1130），金兵南侵，端躬与父大理寺评事若钧、伯中奉大夫传（原名若古）、兄衍圣公端友等护驾南渡。伯与兄寓居衢州，端躬则侍父随驾，抵台州之章安镇。时端躬目睹朝廷腐败，奸臣揽权，叹息不已，自以枉为朝官，意欲丢官弃禄，做一庶民，觅栖身之地，自食其力，与草木为邻。道经永康榉川（今金华市磐安县盘峰乡榉溪村），值父若钧病逝，遂葬父于钟山之后坞，并隐居于榉溪，承先志建山庄古寨"南宗厥里"。宝祐二年（1254），理宗追端躬功德，以衢州孔庙例建榉川南宗厥里孔氏家庙，赐"万世师表"金匾一块，后遂为孔氏婺州南宗。其子孙照例免赋税劳役，白身最长者可荐朝录用。

Kong Duangong（？—1138）had his courtesy name as Zijing. His ancestral home was located at Jueli, Qufu, Shandong. He was the 48th generation of Confucius, and hardworking when he was young and became a successful candidate in imperial examination. At the third year of Song Xuanhe reign（1121）, he acted as Chengshilang and appraised state affairs in Dali Temple. He was clear-handed, suppressed the evil and pacified the good. His peer officials were in fear of his dignified composure, but people expressed thanks to his virtue. At the fourth year of Jianyan reign（1130）, soldiers of Jin regime conducted a crusade into the south, Duangong and his father Ruojun, an appraiser of Dali Temple, his uncle Zhongfeng official Zhuan（original name was Ruogu）and his brother Kong Duanyou together protected the Emperor to go to the south. His uncle and brother lived in Quzhou, Duangong and his father continued the journey along with the Emperor, finally arrived at Zhang'an town of Taizhou. At the moment, Duangong saw the corruption of the imperial court and treacherous court officials usurping power. He sighed for this, and self-thought he was unable to be called an official. So, he wanted to quit his position and became a common person. He found a place for settlement and earned his own living as well as made neighborhood with plants. When they passed by Juzhou, Yongkang（now Juxi village, Panfeng, Pan'an, Jinhua）, his father Ruojun died of illness, so he buried his father at Houwu, Zhongshan. Thereafter, he lived in seclusion at Juxi and inherited ancestral ambition to build a villa and an ancient village "South Jueli". At the second year of Baoyou reign（1254）, Emperor Lizong built Kongs' family temple of South Jueli at Juzhou, in order to memorize the merits and virtues of Duangong, based on the Confucian temple of Quzhou. And he also bestowed this temple a golden tablet,

"Exemplary Character for Generations", making here South Confucius Temple of Wuzhou. His offspring also were exempted from taxation and penal servitude. And, the person who was not holding any government official post for the longest period can be recommended to the court and had a position.

2. 梓誉村蔡氏　The Cais of Ziyu village

梓誉村原名安仁里，"梓誉"两字是从古文"桑梓誉重"一语中摘选出来的。梓誉有着深厚的历史文化底蕴，曾先后有过 4 位岁进士、11 位庠生、18 位太学生。蔡氏宗祠，位于浙江省金华市磐安县双溪乡梓誉村。自 1196 年蔡元定之子蔡渊携其子蔡浩因避祸入居梓誉村后，繁衍后代，至今已有 800 多年历史。蔡元定及其子蔡渊具为蔡氏九儒，研究现学著及之说，深得宋理学家朱熹的好评。因此朱熹赠墨宝"理学名宗"的崇高荣誉。

The original name of Ziyu village was Anrenli. The words, "梓誉" were picked from the ancient phrase "桑梓誉重 (be hardworking and glorify the ancestors)". Ziyu enjoys rich historical and cultural connotation, having four tribute students, 11 students and 18 imperial college students. The Cais' ancestral hall was located as Ziyu village, Shuangxi, Pan'an, Jinhua, Zhejiang. It had passed over 800 years since Cai Yuan, the son of Cai Yuanding, with his son Cai Hao settled here in order to avoid wars and multiplied in 1196. Cai Yuanding and his son Cai Yuan, two of nine scholars of the Cais, studied famous subjects and wrote numerous books, obtaining high praise from Zhu Xi, neo-Confucianist of the Song Dynasty. Hence, Zhu Xi bestowed them his calligraphy work, "famous family of neo-Confucianism."

蔡氏家训　The Cais' Family Instructions

爱国爱家，报效祖国为己任，为官者（国家干部）更应清正廉明，不贪不赃，造福黎民。

Love the country and family. Take serving the country as the duty. Officials must be clear-handed and fair, and bring benefits to people.

尊老爱幼，社会之美德，孝顺父母，爱护少年儿童，责无旁贷。

Respecting the aged and taking good care of children are social virtues. It is our responsibility to be filial to parents, love and protect teenagers.

蔡氏宗祠 The Cais' Ancestral Hall

古民居建筑 Ancient civil building

宗祠内匾额"理学名宗" The tablet in the ancestral hall "理学名宗（Famous Family of Neo-Confucianism）"

兄弟姐妹、夫妻、妯娌、宗亲之间，互相关心，和睦团结，同舟共济，患难与共。

Brothers, sisters, couples, sisters-in-law and relatives should care each other, unify one another, and pull together in times of trouble and share weal and woe.

遵纪守法，崇尚精神文明。待人以礼，遇事晓之以理，不偏不倚，公平合理，一切浮靡斗狠恶习不可沾染。

Observe disciplines and obey laws, advocate mental civilization. Treat people with courtesy; enlighten people with reason fairly and reasonably. Refuse all bad habits.

勤俭创业，发奋读书，造福桑梓。

Be hardworking and thrifty when starting a business, studious when learning, and bring benefits to later generations.

切不可流于娼盗之行列，自贱其身，玷污祖宗。

Please don't be in the rank of prostitutes and thieves, treading on yourself and tarnishing your ancestors.

宗族之盛衰，富贵贫贱，贤愚世代皆有，扶弱济困人之美德，既同源流，不可欺凌。

Every generation will experience the ups and downs of the family, with the rich and the poor, the wise and the silly. Helping the weak and the poor is the moral virtue. Since you all share one origin, you cannot bully each other.

蔡氏子孙，须遵祖训，以理学道德规范量身，树立美好之形象，以为人之表率。

The offspring of the Cais must obey ancestral instructions and require themselves according to neo-Confucianism moral regulations to build a good image and become the example of others.

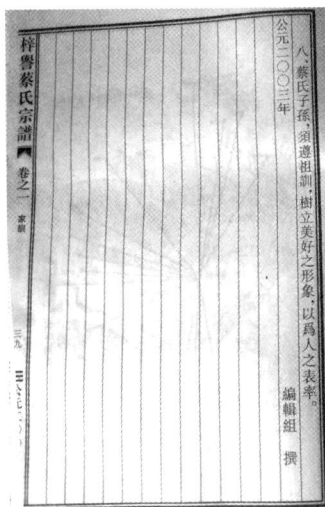

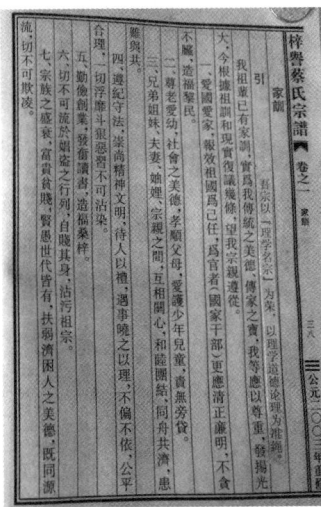

蔡氏家训 The Cai's Family Instructions

磐安县双溪乡梓誉村蔡氏是南宋理学家蔡元定的后裔，蔡氏定居梓誉村已有 800 多年历史。梓誉蔡氏以理学的道德伦理教育后人，其家训是以理学的"仁"和"和"的道德规范为基准制定的，思想理念已深深流入蔡氏子孙的血液里。蔡氏宗谱记载了很多家训故事。

The Cais of Ziyu village, Shuangxi, Pan'an were the descendants of Cai Yuanding, neo-Confucianist of the Southern Song Dynasty. The Cais had settled down there for over 800 years. The Cais of Ziyu educated later generations in line with neo-Confucianism's moral and ethical values. Their family instructions were formulated based on the moral regulation of "benevolence" and "harmony" in neo-Confucianism, and such mental thoughts had deeply permeated in the blood of the Cais' offspring. The Cais' genealogy recorded many stories about family instructions.

蔡承烈　Cai Chenglie

蔡承烈见大溪上平桥被山洪冲垮。母子议建大桥，母当即出私蓄银二百两，三年大桥建成。承烈天天在建桥工地，因劳累过度，时年三十六岁英年而逝。承烈公之功德，族人为刻碑石以记。体现孝亲的有：康熙年代，蔡伯明其母蔡徐氏，夫早卒。蔡徐氏自奉公婆，苦守遗腹子，训以成人。蔡徐氏老成残，在床八年多。蔡伯明及妻，事奉其母，躬亲溺浣，扶起晨昏，夜以继日，衣不解带，不安枕，奉养无怠。

Cai Chenglie saw that the Ping bridge above the big stream was burst by torrential flood. He and his mother suggested rebuilding the bridge. His mother immediately gave out her personal savings, 200 liang of silver. This bridge was completed within three years. Chenglie spent his days at the construction site and died at 36 due to excessive fatigue. The merits and virtues of Chenglie was engraved on the gravestone by his family members. The story showing filial piety was: during the reign of Kangxi, the mother of Cai Boming, Mrs Xu of Cai lost her husband early. She served her parents-in-law, and sadly stayed with her son and educated him to be a real man. When she was old, she became disabled and laid in bed for over eight years. Cai Boming and his wife, served his mother, personally washed her nasty clothes, helped her get up at dawn and dusk. Day after day, they were often sleepless to support and wait upon their mother, without sluggish moments.

五、浦江 Pujiang

郑宅镇郑氏 The Zhengs of Zhengzhai town

郑义门，又称"江南第一家"，位于浙江省金华市浦江县郑宅镇，占地约5000平方米，是中国古代家族文化的重要遗址。自北宋崇和元年（1118）至明天顺三年（1459），郑氏家族在此合族同居历时350余年，以孝义治家闻名于世。长达168条的传世家训《郑氏规范》，被誉为中国传统家训的重要里程碑。其事载入《宋史》《元史》《明史》。今天的郑义门，是全国重点文物保护单位、浙江省廉政教育基地以及爱国主义教育基地。

Zhengyimen, also "The Best Family of Jiangnan", was located at Zhengzhai town, Pujiang county, Jinhua, Zhejiang province. Covering about 5,000 square meters, it was an important site of ancient Chinese family culture. From the first year of Northern Song Chonghe reign(1118) to the third year of Ming Tianshun reign(1459), the Zhengs had lived there for over 350 years, well-known for its governance over the family based on the devotion to parents and the loyalty to friends. Extant family instructions, *The Zhengs' Family Regulations*, with 168 articles, is reputed as a vital milestone of traditional Chinese family instructions. Their family events were recorded in *History of Song Dynasty*, *History of Yuan Dynasty* and *History of Ming Dynasty*. Zhengyimen today is a state-level key cultural relics protection unit, anti-corruption education base and patriotism education base of Zhejiang province.

郑氏宗祠坐东朝西，占地约5000平方米，共分为5进64间，其建筑无雕梁画栋之华丽，保简朴整洁之本质，庄严宽敞，古朴厚重。走进宗祠，就像走进了中国历史的博物馆。内有元丞相脱脱书写的"白麟溪"碑，明皇帝朱元璋亲赐的"江南第一家"牌匾，明代"开国文臣之首"宋濂手植的苍劲古柏，范仲淹、朱熹、柳贯、王锡爵等历代历史名人及当代书法家题写的大量匾额、楹联等……正可谓翰墨丹青赏心悦目，诗书雅乐神韵悠远。让今天的人们依然津津乐道的，是郑氏家族魅力不减的人文遗风。

江南第一家 **The Best Family of Jiangnan**

Sitting toward the east, Zhengs' ancestral temple had coverage of 5,000 square meters. It consisted of five rows, with 64 rooms all together. Without splendid painted pillars and carved beams, this building kept its nature of simplicity and cleanliness, presenting a solemn, spacious, primitive and thick style. When stepping into this temple, it's like entering a Chinese historical museum. Inside there is a "Bai Lin Xi" tablet written by the Prime Minister of the Yuan Dynasty, Tuo Tuo, "The Best Family in Jiangnan" plaque bestowed by Zhu Yuanzhang, the Emperor of the Ming Dynasty, a row of vigorous ancient cypresses planted by Song Lian, the "first founding civilian official" of the Ming Dynasty, and a great amount of tablets and couplets written by historic celebrities and modern calligraphers, like Fan Zhongyan, Zhuxi, Liu Guan and Wang Xijue. These calligraphy works and paintings were feasts to eyes, while poems, books, music were expressing a remote romantic charm. It was the enchanting cultural heritage left behind by the Zhengs that people took delight in talking about today.

《郑氏规范》中治家、教子、修身、处世的家规族训，以及极具特色的教化实践，对中国古代家族制度的巩固发展，对中国封建社会后期的稳定和儒家伦理、文化的世俗化，都产生了深远的影响。朱元璋看重郑氏家族孝义治家，耕读为本的家规家法，在明代的法律中引入了不少《郑氏规范》的内容。浦江孝义门郑氏历经宋、元、明三代十五世，同居共食达350年，最多的时候，有3000人。

"忠孝传家" 匾 The tablet inscribed with "忠孝传家 (Passing Down Loyalty and Filial Piety)"

门口题词 The inscription on the entrance

In *The Zhengs' Family Regulations*, family rules and instructions about family regulation, children education, moral character cultivation and social communication and distinctive humanization practices posed profound influences over the consolidation of ancient Chinese family system, and the stabilization of later-stage Chinese feudal society and the secularization of Confucian ethics and culture. Zhu Yuanzhang thought highly of the Zhengs' family rules and regulations of governing the family by filial piety and part-work and part-study lifestyle, so he cited many contents of *The Zhengs' Regulations* when formulating the laws of the Ming Dynasty. The Zhengs of Xiaoyimen, Pujiang had witnessed three dynasties and passed down 15 generations. They had lived and eaten together for 350 years.

And the largest number of people living and eating together reached 3,000.

《郑氏规范》规定：郑氏子弟，8 岁入小学，16 岁入大学，能背四书五经，并能讲出正文大义，才允许加冠，成为成人。子弟已冠而习学者，每月十日一轮，要考查经文。明代开国大臣宋濂，曾在郑氏执教 20 余年，为郑氏培养了许多人才。

The Zhengs' Family Regulations regulated that: the Zhengs' offspring must go to primary school at 8 years old and to college at 16 years old. They must be able to recite the Four Books and the Five Classics, and can tell the general meanings. Only achieving these can they be allowed to be crowned and be an adult. As for those adult learners, they will be tested to recite the classics once in ten days. Song Lian, the founding official of the Ming Dynasty, had taught for over 20 years in the Zhengs' family, cultivating many talents for this family.

规范还规定：子孙出仕有以赃墨闻者，生者则于《谱图》上削去其名，死则不许入祠堂。从宋到元，郑氏已有多人为官，在明代，共有 47 人为官，官位最高的是礼部尚书。令人惊叹的是，郑氏子孙没有一个因贪污被罢官。

It also regulated that: if the offspring became officials but committed corruption, their name would be removed out of the *Genealogy* if they were alive, or they were not allowed to enter the hall if they died. From the Song Dynasty to the Yuan Dynasty, there were many officials in this family. And in the Ming Dynasty, there were 47 officials, among which the highest official was director of the Board of Rites. Surprisingly none of them was dismissed because of bribes and corruption.

第十一条，"毋徇私以妨大义，毋怠惰以荒厥事，毋纵奢侈以干天刑，毋用妇言间和气，毋为横非以扰门庭，毋耽曲蘖以乱厥性……"第十八条，"子孙赌博无赖及一应违于礼法之事，家长度其不可容，会众罚拜以愧之……"第五十九条，"子孙以理财为务者，若沉迷酒色、妄肆费用以致亏陷，家长复实罪之……"第六十五条，"亲姻馈送，一年一度，非常吊庆则不拘。此切不可过奢……"第八十六条、第八十七条讲到，"……即仕，须奉公勤政，毋踏贪黩，以忝家法，……""……当早夜切切以报国为务。抚恤下民，实如慈母之保赤子；有申理者，哀矜恳恻，务得其情，毋行苛虐。有不可一毫妄取于民。"

第一百一十三条，"子孙不得从事交结，以保助闾里为名而恣行己意，遂致轻冒刑宪……"第一百二十五条"子孙不得惑于邪说，溺于淫祀，以邀福于鬼神。"

The 11th article, "Do not harm the righteous cause because of private affairs. Do not get slack at your business because of your laziness. Do not indulge and be punished by the heaven. Do not harm the harmony of the big family because of your wife's slanderous talk. Do not do evil to degrade the family. Do not indulge in wine and food and forget a man's real nature." The 18th article, "If the descendants do gambling or other violation of the law, the patriarch should punish them in the face of public according to the severity of bad deeds. They will be made to bow to the elders and feel shameful." The 59th article, "Suppose the descendants in charge of finance are indulged in booze and women, or in reckless spendthrift, leading to operating losses and missing accounts. Once verified by the patriarch, they have to face the same punishment…" The 65th article, "Bestow gifts to clan members once a year, unless there are special occasions like weddings and funerals. Such occasions should not be too luxurious." The 86th article, "After they have become government officials, they should respect justice and abide by the laws, work hard on government affairs." The 87th article, "should always bear in mind how to repay the state, understand and sympathize with the poor and take care of them just like a mother. They should hold a heart of compassion for those with wrong charges, find out the truth and shall not abuse them. They should not take anything from the people." The 113th article, "Descendants should not gang up in village and act recklessly in the name of protecting the townsmen, so much so that it violates the laws and decrees of the country…" The 125th article, "Descendants should not be bewitched by superstition and heresy for blessings. They should not get obsessed with messy sacrifices not in accordance with social rites."

郑氏家规共计 168 条，汇聚了郑氏家族数代人思想道德素质的精华，其中大多数与我们现在倡导的"八荣八耻"可以相辅相成。

The Zhengs' family rules, altogether 168 articles, collected the essence of ideological and ethical standards of generations of family members of the Zhengs, most of which can supplement "Eight Honors and Eight Disgraces" advocated by people now.

取义成仁郑洧　Zheng Wei sacrificed life for justice

取义成仁是郑义门的一个故事。明朝时监生私纳赃款，篡改丈量田地的鱼鳞图册，导致当时担任一方粮长的浦江郑氏家族的族长郑濂因失察而获连坐之罪。刑部差人抓走郑濂，另外几个兄弟争相要入京替哥哥承罪，争来争去，最小的弟弟郑洧力排众议只身来到南京。兄弟争相入狱被传为佳话，这事情感动了朱元璋，他不但没有治罪郑家，反而给郑洧升了官。朱元璋当时就对郑濂说，"你家九世同住，孝义名冠天下，果然名不虚传，可谓天下第一家"。这很好地诠释了孝忠两全。凭借好学的风尚和孝义的名声，从宋、元到明、清，郑义门约有173人为官，尤其是明代，出仕者达47人，官位最高者位居礼部尚书。令人惊叹的是，郑氏子孙中，竟没有一人因贪被罢官。

Sacrifice life for justice, one story of Zhengyimen. During the Ming Dynasty, the students of the Imperial College personally took in illicit money and falsified land registration books that used to measure the farmland, thus resulting in punishing Zheng Lian due to his oversight. At the moment, Zheng Lian, the Patriarch of the Pujiang Zhengs, acted as the official responsible for grain. The Ministry of Punishments sent official to arrest Zheng Lian. The brothers of Zheng Lian competed to take the criminal liability of their elder brother. After long quarrel, the youngest brother, Zheng Wei came to Nanjing alone. Such thing became a favourite tale, moving Zhu Yuanzhang. He did not punish the Zhengs, but gave Zheng Wei a promotion. At the moment, Zhu Yuanzhang said to Zheng Lian, "your nine generations lived together, and your filial piety was well-known across the country. You really deserved the reputation, the best family in the world." It perfectly explained that serving the Emperor with loyalty and your parents with filial piety simultaneously. On account of studious lifestyle and filial reputation, Zhengyimen had cultivated 173 officials from the Song, Yuan to Ming and Qing Dynasties, 47 officials in the Ming Dynasty in particular, and the highest official position was the Minister of the Ministry of Rites. Surprisingly, there was no one quitting his position due to corruption in the offspring of the Zhengs.

六、武义 Wuyi

1. 明招山吕氏 The Lvs of Mingzhao Mountain

吕祖谦所属家族东莱吕氏是一个延续了百多年的大家族，人才辈出，究其原因，正在于家规家训的教化。吕氏家族早在北宋吕夷简时就作有《门铭》。吕祖谦有《闰范》《少仪外传》《家范》《辨志录》等。《家范》收录在吕祖谦的《东莱集》中，共有六卷，分别为《宗法》《昏礼》《葬仪》《祭礼》《学规》《官箴》，从敬宗收族、明理躬行、清慎勤实等方面阐述了其家训思想。《宋史》赞之为"居家之政，皆可为后世法。"

The family, the Lvs of Donglai, where Lv Zuqian belonged was a big family of over one hundred years. In this family, men of talent come out in succession due to the moralization of family rules and instructions. As early as the Northern Song Dynasty when Lv Yijian was alive, this family wrote out *Men Ming*. Lv Zuqian wrote *Kun Fan*, *Shaoyi Wai Zhuan*, *Family Rules* and *Bian Zhi Lu*. *Family Rules* was collected in Lv Zuqian's *Donglai Collection*. It had six volumes, including *Family Regulations*, *Wedding*, *Funeral*, *Sacrifice*, *Student Rules* and *Warnings for Official*. It elaborated his family instruction ideology from family respect and unification, reasoning and practicing, honesty and hard work. *History of the Song Dynasty* commented it, "the way this family governed its people can be an example for later generations."

明招讲院内的传薪亭 Torch Relay Pavilion inside of Mingzhao Academy

明招讲院一隅 A corner of Mingzhao Academy

明招讲院内景 Internal scene of Mingzhao Academy

吕氏家范（摘编）Family Regulations of the Lvs（excerpts）

敬宗祖 Respect ancestors

亲亲故尊祖，尊祖故敬宗。此一篇之纲目。人爱其父母，则必推其生我父母者，祖也。又推而上之，求其生我祖者，则又曾祖也。尊其所自来，则敬宗。儒者之道，必始于亲。此非是人安排，盖天之生物，使之一本，天使之也。譬如木根，枝叶繁盛，而所本者只是一根。

Only loving parents can we respect ancestors, thus admire our patriarchal clan. This is the outline of this essay. The love for our parents helps us to trace the people that gave birth to our parents, namely grandparents, then the people that generated our grandparents, our great-grandparents. Respecting our origin requires us to admire our patriarchal clan. The doctrine of scholars must start from our filial piety with our family. This is not an artificial management. This is because the heaven gives life to all things, making them a same origin. Such as tree roots, despite of the flourishing leaves, they have only one root.

严治家 Manage the family strictly

子弟不奉家庙，未冠执事很慢，已冠颓废先业，并行榎楚。执事很慢，谓祭祀时醉酒，高声喧笑，斗争，久待不至之类。颓废先业，谓不孝、不忠、不廉、不洁之类。凡可以破坏门户者，皆为不孝。凡出仕，不问官职大小，蠹国害民者，皆为不忠。凡法令所载赃罪，皆为不廉。凡法令所载滥罪，皆为不洁。

If the offspring didn't worship the family temple, those under 20 neglected works, while those above 20 made ancestral career decline, they would be punished. The former one refers to getting drunk when doing worship ceremony, laughing aloud and having fights and not arriving while others wait for long time. And the latter one refers to impiety, disloyal, corrupted and unclean. The person who harms the family fame was not filial. If being an official, no matter high-rank or low-rank, the person who hurts the country and people is disloyal. All embezzlement and bribe-taking crimes recorded by the laws are corrupted deeds. All misuses of authority recorded by the laws are unclean behaviors.

重品性 Attach importance to moral character

凡预此集者，闻善相告，闻过相警，患难相恤，游居必以齿，相呼不以丈，不以爵，不以尔汝。

For all students who come to learn, they should inform each other when hearing good things, warn each other when hearing others' fault, help each other when having troubles, go out and back, sleep and wake up according to ages, call each other without using title of honor or nobility, you and me.

育实材 Cultivate true talents

凡与此学者，以讲求经旨、明理躬行为本。肄业当有常，日纪所习于簿，多寡随意。如遇有干辍业，亦书于簿。一岁无过百日，过百日者，同志共摈之。凡有所疑，专置册记录。同志异时相会，各出所习及所疑，互相商榷，仍手书名于册后。怠惰苟且，虽漫应课程，而全疏略无叙者，同志共摈之。不修士检，乡论不齿者，同志共摈之。

——摘自吕祖谦《家范·宗法》

For those who come to study, they should focus on pursuing the main idea of classics and understand the truth and practice. The learning should be regulated. Every day, freely write down what you learned on the notebook. If you do have to stop studying due to something, you should write down the reason. The days without learning should not exceed 100 days. If it exceeds more than 100 days, this person would be despised. The knowledge that makes you puzzled should be recorded on a special notebook. When meeting each other, you can discuss with others on what you learned and what you have trouble with. You should also write your name on the back of the notebook. For those lazy students, even though he casually deals with the class, he would make a rough notebook without orders. He ought to be despised by the people. If a person doesn't cultivate the virtue of a scholar to make local people feel shameful, the people should despise him together.

——From Lv Zuqian's *Family Regulations · Patriarchal Rules*

倡清廉 Advocate incorruptibility

当官之法，唯有三事：曰清，曰慎，曰勤。知此三者，则知所以持身矣。

当官处事，常思有以及人。

——摘自吕本中《舍人官箴》

There are three rules for being an official: incorruptible, cautious and diligent. Knowing these three rules, you will know how to self-cultivate.

Being an official, you should often think of how to bring benefits to the people.

——From Lv Benzhong's *An Official's Expostulation to Other Officials*

吕祖谦　Lv Zuqian

吕祖谦出生于 1137 年，出生于婺州。他的祖上八代曾出过 17 位进士、5 位宰相。高祖吕希哲（1039—1116），自其以下，包括曾祖父吕好问、伯祖父吕本中、祖父吕弸中、父吕大器等，皆为朝廷命官。如此一代一代连续不断地在朝为官，在宋朝前后的朝代中不多见。

Lv Zuqian was born in 1137 at Wuzhou. Among his former eight generations, there were 17 successful candidates in the imperial examination and 5 Prime Ministers. His great-great-grandfather was Lv Xizhe

吕祖谦 Lv Zuqian

（1039—1116）, great-grandfather Lv Haowen, grandfather's elder brother Lv Benzhong, grandfather Lv Pengzhong and father Lv Daqi were all appointed officials. Such a family having appointed officials in continuous generations is not commonly-seen in the dynasties either before or after the Song Dynasty.

吕祖谦少儿时期曾经跟随为官的父亲在福建生活，他曾经师从林之奇、汪应辰、胡宪游。这几位都是宋朝的文学家。后面师从伯祖吕本中。当时，学人多称其伯祖吕本中（1084—1145）为"东莱先生"，吕祖谦则被称为"小东莱先生"。到了后世，一般均称吕祖谦为"东莱先生"了。学业上吕祖谦受吕本中的影响最大。1161 年，吕祖谦 25 岁，他的祖父吕㻏中想让他为官，让他任严州桐庐县尉，主管学事。但是他没有去上任，偏偏要去参加科举考试。两年后，吕祖谦考中博学宏词科，接着又中进士。因为两门都中了，朝廷也特授他左从政郎（八品官员）。但是时运不济。在吕祖谦中进士的前一年，他妻子去世，儿子夭折。4 年后，母亲去世，下葬于婺州，吕祖谦为母亲守丧，只能留在婺州教书。1169 年，也就是离母亲去世 3 年后，他再娶韩氏（为原配之妹），并到严州任所。干道六年（1170），他升任太学博士，并兼国史院编修官。但是次年，妻子韩氏去世，女儿夭折；次年（1172）他的父亲去世。他又为父亲守丧 3 年。这三年他依然以教书写著作为主。淳熙三年（1176），守丧期满，因李焘的推荐，升任秘书省秘书郎，并兼国史院编修官与实录院检讨官。这一年，他 40 岁，身体健康状况已经大不如前，疾病缠身。淳熙四年（1177），他娶芮氏为妻，2 年后其妻芮氏去世，这一年他 42 岁。1181 年，吕祖谦病故，享年 45 岁。

　　When Lv Zuqian was young, he lived with his father who worked in Fujian. And he learned from Lin Zhiqi, Wang Yingchen and Hu Xianyou, all men of letters of the Song Dynasty. Later on, he learned from his grandfather's elder brother Lv Benzhong. At the moment, his grand-uncle was reputed as "Donglai scholar", while Lv Zuqian was called as "Donglai scholar junior". Later generations generally called Lv Zuqian as "Donglai scholar". Lv Benzhong posed the greatest influences over Lv Zuqian's academic career. In 1161, Lv Zuqian was 25 years old. His grandfather Lv Pengzhong wanted to appoint him an official position, an official that was just below the head of the county of Tonglu, Yanzhou, mainly charging learning. But he refused this appointment and deliberately jointed imperial examination. Two years later, he passed extensive learning and great literary talent test and then passed the imperial examination. For his success in these two exams, the imperial court specially appointed him as left governmental official (eight-class official). But out of luck, his wife died and his son died young in a year before Lv Zuqian passed the imperial examination. Four years later, his mother died and was buried in Wuzhou. He kept vigil beside the grave of his mother and could only stay at Wuzhou to teach students. In 1169, namely three years after his mother's death, he remarried Mrs Han (the younger sister of his first wife) and worked at Yanzhou. At the sixth year of Gandao reign (1170), he was promoted to be court academician of Imperial College and compiler of National History Academy. But next year, his wife, Mrs Han and his daughter died. Next year (1172), his father died. He kept vigil beside the grave of his father for three years, during which he still taught students and wrote books for living. At the third year of the Chunxi reign (1176), he finished his wake. Recommended by Li Tao, he was promoted to be a secretary of the secretariat and compiler of National History Academy as well as examiner of Memoir Academy. This year, he was 40 years old. His physical condition was far worse than before and clouded with a series of illnesses. At the forth year of the Chunxi reign (1177), he got married with Mrs Rui. Two years later, his wife passed away when he was 42 years old. In 1181, he died of illness at the age of 45.

吕祖谦墓 Tomb of Lv Zuqian

2. 后陈村陈氏　The Chens of Houchen village

据《陈氏宗谱》记载，明成化年间，兵戈扰乱，农业荒歉，陈之谟（富八公）由义乌县双林乡下园西陈村迁至武义城东4公里陈高山西居住。后为方便耕种，又从陈高山移到皮店东侧，称后陈村。

According to *The Chens' Genealogy*, during the Chenghua reign of the Ming Dynasty, the confusion and disorder brought about by war led to crop failure, forcing Chen Zhimo (Fubagong) to move to the west of Chengao mountain, 4 kilometers away from the east of Wuyi from Xichen village, Xiangxiayuan, Shuanglin, Yiwu. Later, for convenient farming, he moved to the east of Pidian from Chengao mountain, called Houchen village.

夫双林诚义邑之名区，为婺郡之胜地，向因有明而至成化年间，兵戈扰乱兼之荒歉，频仍故里尽遭回禄乃至。富八公讳之模而至武邑，行至陈高山之西，乃见四山环绕二水会合，卜筑于后陈迁居焉，郡曰颍川其前代流传所志，虽不能详悉而于始祖之所，自出原原本本悉有自来，故谨以备载之。

This place was a famous district of Chengyi county, Shuanglin, a well-known scenic spot of Wu county. However, the confusion and disorder brought about by war led to crop failure, combined with continuous conflagration in this native place, leading to such scene for the moment. Fubagong, with his name as Zhimo, arrived at Wuyi and walked to the west of Chengao mountain. He saw two

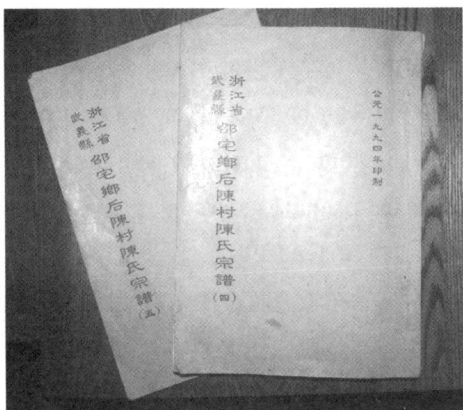

陈氏宗谱*The Chen's Genealogy*

rivers joined together and four mountains embraced this place, so he built his house in Houchen village. This legend was passed down by the previous generation of Yingchuan, and people didn't know the place of ancestor but they knew this legend must have a origin. Hence, they recorded this story.

后陈陈氏家训（摘选）　Family Instructions of the Chens, Houchen (excerpts)

孝　Filial piety

夫孝为百行之原，彝伦之首，人自有生以来，身体发肤受之父母，诗云抚我畜我长我育我欲报之德，昊天罔极，为人子者，顾可一念忘其亲乎！是故黄香扇枕孟宗哭竹，发诸天性，彼虽幼而已然，谁非人子乎哉！自当内尽其心，外竭其力，务使制行不亏，身无玷缺，贫者纯艺黍稷，牵车服贾，以孝养厥父母，富者奋志，诗书显亲，扬名以报亲恩于万一。

Filial piety was the origin of all behaviors, and the first logical thinking. Since a person was born in this world, his body, skin and hair was given by his parents. Poems often said like this, the parents cultivated, fed and educated me, and their grace was so great that I had no way to repay. Being the child, I cannot forget their grace of cultivating me. So Huang Xiang cooled the mattress with a fan in hot summer and warmed the quilt with his body in cold winter for his father, and Meng Zong cried in the bamboo woods for methods to save his mother. Their behaviors both came from their nature. They were so young but also did such things. Everyone was the child. So we should try our best mentally and physically not to destroy regulated rules and norms and have no flaws. The poor should focus on farming, ride bull cart and do business, in order to serve parents with best filial piety. The rich should cheer up and make his parents well-known through his studying to pay back the parental grace.

友　Friendship

人之有兄弟犹身之有手足，同气之宜总由一本。《小雅》云：脊令在原，兄弟急难。每有良朋，况也永叹。则知凡今之人莫如兄弟。夫年长以倍则父事之，十年以长则兄事之，况同气之人乎且让梨让枣，天良发见于幼穉，谁非兄弟，顾可以阋墙召变致乖骨肉乎，俗有云难得者兄弟，易得者田地，田地务须，饮食必逊，语言必顺，步趋必徐，行坐立必居下式，好无尤兄弟怡怡以尽壎篪之雅。

The brother to a person was like the legs to a body, of the same breath and branches. *Xiao Ya* said, brothers should help each other when meeting with harsh difficulties. Even having some good friends, what they can do was only to comfort, making people sigh that, the best friend in the world was his own brother. If the brother's age was several times of yours, you serve him as a father; within ten years, you serve him as a brother. Brothers would give away bigger pears or Chinese dates. Their natural kindness was presented in their childhood. For brothers, it was not good to have quarrels and harm the relationship. It was often said that, the relationship of brothers was more important than property. You should manage your farmland, stick to your common food, speak gentle language, walk slowly and humbly, thus making a good brother.

睦　Harmony

内之有宗族也，犹水之有分派，木之有分枝，虽远近异势，疏密异形，要其本原则一礼，日尊祖故敬宗，敬宗故睦族。明人道必以睦族为重也，奈何不思子姓之众，皆出于祖宗一人之身，竟相视为途人而不顾哉，昔张公艺九世同居，江州陈氏七百口共食，凡属一家一姓当念乃祖乃父，宁厚无薄，宁亲无疏，长幼必以序相洽，尊卑必以分相聊，务使血脉相通，病瘝相关。

The family had different branches, just like the water had different branches, and the tree also had different twigs. Although they had various forms or spacing, they all came from one origin. So we said the respect to our ancestors was the respect to our clan, which also harmonized our family. People often said we should attach importance to harmony our family. However, all descendants were created by one ancestor, so it was bad to regard family members as passers-by and ignore them. The nine generations of Zhang Gongyi lived together, and 700 people of the Chens of Jiangzhou ate together. So the people had the same surname should have a good treat on each other, be close to each other. The elders and the young should live in harmony and the superiors and the inferiors should chat according to their statuses, making them be related by blood and show concern for mutual well-being to each other.

东汉陈氏三君　Three gentlemen of the Chens in the Eastern Han Dynasty

元方、季方的父亲名陈寔，家庭贫困，但自幼好学。当上了"太邱长"（后任大将军窦武掾属）。陈寔廉洁奉公，断事公平，百姓很佩服他。陈寔生活非常俭朴，家里不用佣人。长子元方、次子季方兄弟两人都有很高的才学和德行。陈寔外出办事，都叫元方在前面拉车，季方在后面拿着手杖，路上行人见了，无不称赞他们。因为陈寔父子三人为官廉洁贤明，声望极高，史称"东汉陈氏三君"，当时豫州城墙上画着他们父子三人的画像，好让百姓们学他们的品德。

The father of Yuanfang and Jifang was named Chen Shi. Born in a poor family, he was very studious. When serving as "Taiqiuzhang" (later assistant officer of Dou Wu, a senior general). Chen Shi was clean and devoted in his work, and made fair judgment, admired by people. Chen Shi led a thrifty life and never hired servants in his family. His elder son Yuanfang and younger son Jifang both enjoyed good talent and virtue. When Chen Shi went out for business, he would ask Yuanfang to pull the car in the front while Jifang held his cane behind. People on the road saw this scene and all thought highly of their behaviors. On account of their clean and wise governance as well as great reputation, they were called as "three gentlemen of the Chens in the Eastern Han Dynasty". The portrayals of the father and his sons were drawn on the city wall of Yuzhou, in order to encourage people to learn their good virtue.

元方的儿子名叫长文，季方的儿子名叫孝先，有一天各说他俩的父亲的功德高，相互争论起来，争来争去，各不相让。于是二人就请祖父陈寔裁决。陈寔对两个孙子说"元方难为兄，季方难为弟，他俩的功德都很高，实在难以分出上下"。由此，陈寔这番话就成为我国古汉语中一句非常有名的成语，即"难兄难弟"。这里的"难"字读音为 nán，其真正的含义是——功德相等，难分上下，是褒义词。可是现在人们口头语的"难兄难弟"的"难"字读成 nàn。变为同样恶劣，同样处于困境，是贬义词。

The son of Yuanfang was named Changwen, and the son of Jifang was named Xiaoxian. One day, they were discussing whose father's virtue was higher. They had a quarrel and refused to give in to the other. So, they asked their grandfather, Chen Shi to judge. Chen Shi said, "it was difficult for Yuanfang to be a good elder brother, and it was also difficult for Jifang to be a good younger brother. So their virtues were too high to know their place." Hence, what Chen Shi said resulted in a very famous idiom, namely "nan xiong nan di"). Here, the pronunciation of "nan" was in level tone, making the real meaning of this phrase, two persons had equal virtues and it was hard to tell which one was better. This phrase belonged to commendatory terms. However, "nan" in current catch phrase of people, "nan xiong nan di" was pronounced as falling tone. Its meaning now was that two persons were facing the same severe plight, making this phrase a derogatory term.

七、永康　Yongkang

1. 厚吴村吴氏　The Wus of Houwu village

厚吴村隶属浙江省永康市前仓镇，位于东经 120 度 4 分，北纬 28 度 47

分。面积 2.45 平方千米，耕地 2513 亩。距永康市中心 15 千米，距前仓镇政府所在地 2 千米。地处永康市南陲，南与缙云县接壤，北与后郑村比邻，东接前仓村，西连荆州村，县道下前线穿村而过。南溪西来潆洄东北，屏山耸翠拱卫其南，西临沃野旷阔丰庚，是典型的傍山依水的江南丘陵谷地村落。现有1078 户 3000 多人口，是永康市最大的行政村，中国历史文化名村。

Houwu village belongs to Qiancang, Yongkang, Zhejiang. It is located at 28 degrees, 47 points north, 120 degrees, 4 points east. It covers 2.45 square kilometers with 2,513 mu of farmland. It is at a distance of 15 kilometers from the downtown of Yongkang, and 2 kilometers from the government of Qiancang town. It is located at the southern frontier of Yongkang, its south borders on Jingyun county, its south is near Houzheng village, its east borders on Qiancang village, and its west is near Jingzhou village. Pathways through counties crosses the village. The southern stream comes from the west and winds its northeastern part, mountains with tall and green trees protect its southern part, and its western part is close to broad and rich plain, making a typical water-dependent Jiangnan hill valley village. It is the biggest administrative village in Yongkang and a famous Chinese historical and cultural village.

家规家训 Family rules and family instructions

——纲常至大，勿倒行逆施；

——循分谦谨，勿撒波横肆；

——勤耕务本，勿好逸恶劳；

——艺精贾衡，勿荒业欺诈；

——教读崇儒，勿荡检卑污；

——善事父母，勿忤逆翁姑；

——顺从尊长，勿高抗分理；

——尽忠报国，勿因私忘公；

——诚信为本，勿言行不一；

——冠当笃志，勿无所用心；

——婚须门当，勿高攀低就；

——丧循古礼，勿信道听巫；

——葬礼随分，勿延缓耽搁；

——祭应诚敬，勿亵渎先灵；

——立嗣承桃，勿混乱宗支；

——奖贤励节，勿良莠不辨；

——惩恶守法，勿作奸犯科；

——贞静贤能，勿诲淫乱伦；

——急公输粮，勿耽延职责；

——释争社讼，勿争强好胜。

家规家训 Family rules and family instructions

—Three main-stays and five constant virtues are the most important, and don't act against these rules;

—Abide by the office and be modest and cautious, and don't be unreasonable and make a scene;

—Work hard, and don't love ease and hate work;

—Learn the skills and make a balance, and don't abandon work and cheat others;

—Teach and pursue Confucian studies, and don't be rakish or despicable and filthy;

—Attend upon parents, and don't disobey parents-in-law;

—Be obedient to reputed elders, and don't act independently and defiantly;

—Repay the country with supreme loyalty, and don't forget business due to interests;

—Be honest, and don't make words not matched by deeds;

—Be an adult with single-hearted devotion, and don't be indifferent to anything;

—Be well-matched in social and economic status for marriage, and don't pursue a higher position or yield to a lower one;

—The funeral should follow ancient rites, and don't listen to the heresy;

—The funeral should be acted according to the status, and don't delay;

—Be sincere when holding sacrifice, and don't be disrespectful to the ancestors;

—Make sure heirs and inherit the temples that sacrifice the ancestors, and don't disturb family members;

—Award talents and encourage thrifty, and not distinguish good from bad;

—Punish bad people and obey the rules, and don't commit crimes;

—Be chaste and gentle and able and good, and don't carnalize and incest;

—Make earnest efforts to promote public good like transporting grains, and don't delay duties;

—Don't desire to excel over others when having a quarrel.

永康吴姓始祖 The first ancestor of the Wus of Yongkong

厚吴村吴姓始祖名昭卿，字明之，号屏山，生于宋绍兴乙丑年（1145），卒淳祐癸年（1250）。原系仙居银青光禄大夫全智公十世孙，秉性高亢超然，有出尘之想。其时伯叔、兄弟侄辈先后登仕路要津，济济相踵，人皆为荣。公独视之澹泊如常。偕伯父公之永康任承事郎，见永康南乡武平山明水秀，遂啸咏其间，聚庐而托处焉，是永康吴氏之始祖。

The first ancestor of the Wus in Houwu village had his first name as Zhaoqing, courtesy name as Mingzhi, and self-tilted himself as Pingshan. He was born in the 15th year of Shaoxing reign（1145）and died in the third year of Chunyou reign（1250）. He was the 10th generation of Quanzhigong, Guanglu official of Yinqing, Xianju. His character was detached and wanted to be a monk. At the moment, his uncles, brothers and nephews all became officials, regarded as a glory by the people. However, he didn't seek for interest and fame as usual. He went to Yongkang with his uncle who went to there to be Chengshilang. He appreciated the picturesque scenery of Wuping, Nanxiang, Yongkang, so recited poems there and settled down, thus becoming the first ancestor of the Wus of Yongkang.

2. 库下村胡氏　The Hus of Kuxia village

胡则，字子正，永康人。明道元年（1032）八月，当时任工部侍郎、集贤院学士的他因见江南大旱，江淮地区（淮河以南、长江下游一带）尤为严重，饿死了很多百姓，心急如焚，毅然上书向皇帝请求永远免除江南各地的身丁钱，最后皇帝经过考虑诏令永远免除衢州、婺州两州身丁钱。百姓感其恩德，在方岩山顶广慈寺立庙供奉。

Hu Ze had his courtesy name as Zizheng and was born in Yongkang. In 8th month of the first year of Mingdao reign(1032), he acted as an official of the Ministry of Works and a scholar of Mandarin House. He witnessed the great drought in Jiangnan, especially worse in Jianghuai district (southern Huai River and the lower branch of the Yangtze River). Many people were died of starvation. Seeing that, his heart was torn with anxiety, so he firmly submitted a written statement to the Emperor. He begged for permanently exempting the poll taxation of all places of Jiangnan.

胡氏家谱 *The Hus' Genealogy*

At last, the Emperor decided to permanently exempt the poll taxation of Quzhou and Wuzhou. The people wanted to express thanks to him, so they consecrated him by building a temple at Guangci Temple on top of Fangyan Mountain.

胡公墓 Hu Ze's tomb

胡公庙 Hu Ze's temple

胡氏家训　The Hus' Family Instructions

行善篇：家道盛衰。皆系于积善与积恶而已。何谓积善，居家则孝悌，处事则仁恕，凡所以济人者皆是也；何谓积恶，恃己之势以自强，尅人之财以自富，凡所以欺心者皆是也。

Beneficence：The prosperity and decline of a family property are up to do good and accumulate merit or do wrong. The former behavior refers to respect the elders and be friendly to brothers at home and be benevolent and tolerant when doing things. All things that helping others can be called doing good. The latter is that pretending to be strong by one's own power and make oneself rich by plundering others' wealth. All behaviors that cheat oneself is doing wrong.

释义：家业的昌盛和衰败，都取决于是积德行善还是积恶为患。什么是积善，就是在家中要做到尊敬长辈友爱兄弟，立身行事做到仁善宽容，凡是救助他人的就都可以称为积善；什么是积恶，就是仗着自己的势力妄自称强，通过掠夺他人财富使自己富有，凡是欺骗自己内心的就都是积恶。

修身篇：正人君子，澹泊明志。为人应以忠孝仁义为上，当以家园为重；先忧后乐，鞠躬尽瘁。

Character Cultivation：An upright gentlemen should be indifferent to fame and wealth and pursue noble and unsullied goals. Be a man, you should take loyalty, filial piety, benevolence and justice as the highest moral code of conduct and give priority to the country. You should firstly think of difficulties and then enjoy. You should be respectful and cautious and try your best to work.

释义：正人君子，应该淡泊名利而使自己志趣高洁。做人应当以忠、孝、仁、义为最高道德准则，应当以国家为重；要忧患在前享乐在后，要恭敬谨慎，竭尽心力。

勤业篇：族中子弟当各勤生业，士者攻其学，农者力于耕，工者专于艺，商者蓄其货。宜未雨绸缪，毋临渴掘井，居家务其质朴，莫贪意外之财，莫饮过量之酒。

Hard-work：Family members should be diligent and conscientious in their own business. Students must try their best to study, farmers must commit to farming, craftsmen must focus on skills and businessmen must save wealth. Set eyes on the long run. Take precautions and don't start to prepare at the last moment. When managing a family, you should pursue simplicity and thrifty. You cannot cling to windfall profit and don't drink too much.

释义：家族中的子弟应当在各自的行业中勤勤恳恳、兢兢业业，读书人要用心攻读学业，务农的人要致力耕种劳作，工匠要专注技能工艺，商人要积聚资产财富。应当把眼光放长远，做到未雨绸缪、防患未然，不要临阵磨枪、临渴掘井。持家应当勤俭朴素，俭朴素，不贪恋意外之财，不饮过量之酒。

劝学篇：为人者至乐莫如读书，至要莫如教子。子孙虽愚，经书不可不读，即时冥顽，纵有开悟之时。读书志在圣贤，为官心存君国。穷则独善其身，达则兼济天下。

Admonition to study：The happiest thing of human must be reading, while the most important thing is to educate children. Although being retarded, the descendants must read the *Four Books* and the *Five Classics*. Even those silly and stubborn people, they would be reasonable after studying. You should try your best to become a wise man through studying and keep the mind of being loyal to the Emperor and love the country when being officials. When you fail at work, you have to preserve your moral integrity and cultivate personal character; when you succeed, you have to bring benefits to people.

释义：人最快乐的事情应当是读书，最重要的事应当是教育子女。子孙即使愚钝，四书五经是一定要读的，即使是愚昧顽固的人，读了书也会有通达醒悟的时候。读书要以努力成为圣贤之人为志向，做官则要始终保持忠君爱国的思想。不得志时，要洁身自好修养个人品德；得志显达时，要造福天下百姓。

孝悌篇：恩莫大于父母，情莫切于兄弟。万恶淫为首，百善孝为先，子孙敢有忤逆父母，气凌尊长者，亲房不得徇情。

Filial piety and fraternal duty：The grace of parents is the biggest, and the relation with brothers is the closest. Obscenity rank top among all vicious moralities, while filial piety to the elders also ranks top among different benevolent virtues. Family members should not show favouritism to those descendants who are not filial to parents and do not respect the elders.

释义：各种邪恶品行中，淫邪欲念是排在第一位的，而各色各样的仁善品行中，孝顺长辈也是排在第一位的。子孙后代中有不孝顺父母、不尊敬长辈的，家族成员不得徇私偏袒。

胡则 **Hu Ze**

胡则（963—1039），初名厕，字子正，永康胡库人，被百姓称为"胡公大帝"。宋太宗端拱二年（989）登进士。

Hu Ze（963—1039）had the original first name as Ce, courtesy name as Zizheng. Born in Huku, Yongkang, he was reputed as "Hu Emperor" by people. He became a successful candidate at the second year of the reign of Song Taizong（989）.

胡则于宋太宗端拱二年（989）登陈尧叟榜进士，开宋朝八婺科第之先河。及第时宋太宗御笔削去"厂"，赐名为则。二年后释褐，补许州许田尉，至72岁，宋仁宗宝祐元年以尚书部侍郎致仕。在他浮沉宦海的47年中，先后出知浔州、桐庐、睦州、温州、信州、福州、杭州、池州、陈州等州郡，按察江淮、京西、广西、陕西等六路使节，并曾担任权三司使吏部流内铨、工部侍郎、兵部侍郎等朝迁重臣。逮事三朝，十握州符，六持使节，选曹计省，历践要途，是北宋前期政坛一位中高级官吏。胡则在政治上力主宽刑薄赋，兴革使民，勤政廉政，做了许多功国利民的好事。

胡则 Hu Ze

At the second year of the reign of Song Taizong（989）, his name was posted on the announcement of Chen Yaosou, first candidate in eight counties of Wuzhou in the Song Dynasty. Emperor Song Taizong personally changed his first name. In two years, he was conferred to be Xutian officer of Xuzhou until he was 72 years old. During the 47 years of being an official, he governed Xunzhou, Tonglu, Muzhou, Wenzhou, Xinzhou, Fuzhou, Hangzhou, Chizhou, Chenzhou, and supervised the ambassadors of Jianghuai, Jingxi, Guangxi and Shaanxi. He also acted as important official like, temporarily Liuneiquan of three offices, Shilang of the Ministry of Works and Military. Experiencing three dynasties, governing ten counties, being ambassador for six times, charging official selection and finance, he was a senior and middle official in the political circle of the early Northern Song

Dynasty. At the first year of the reign of Song Renzong, he quit his job as Shilang of the Ministry of Books. Hu Ze had contributed a lot to the country and people by advocating loose punishment and lesser taxes in politics, encouraging reform to benefit the common people, and proposing a diligent and clean government.

3. 程氏　The Chengs

永康程氏始祖程楷，字公范，自元季迁永以后，至明正德乙亥（1515）已建有"闰三公祠"，龛前正匾为四川按察使程锉所书。坐落于寿山口螺蛳塘山，闰三程权公墓左（今刘英烈士陵园纪念馆）。据《程氏宗谱》记载，当时程氏有男丁410人，女197人，是谓名门望族。

The first ancestor of the Chengs of Yongkang, Cheng Kai had his courtesy name as Gongfan. Since the first settlement in Yongkang in the Yuan Dynasty, "闰三公祠(Run San Gong Hall)" had been built at the reign of Zhengde in the Ming Dynasty(1515). The tablet in the front of the shrine was written by Cheng Ji, judicial commissioner of Sichuan. It was located at Luoshitang Mountain, Shoushankou, on the left side of the grave of Run San, Chengquan(now Liu Ying memorial of martyrs cemetery). According to *The Chengs' Genealogy*, the Chengs' family had 410 men and 197 women, making a real established family.

程氏家规　Chengs' Family Instructions

吾之子嗣：（一）莫不孝二亲。（二）莫弃本逐末，背毁师长。（三）莫盗贼累耻先灵。（四）畏四知，为人仁义。（五）远五刑，莫犯刑戮。（六）政宽以调民。（七）躬事廉俭，敦厚自裕。（八）勤习经艺，引文自饬。（九）用九思立德。（十）无忘好善。

The descendants of the Chengs must do: first, be filial to parents; second, don't be ungrateful to defame teachers and the elders; third, don't be thieves to humiliate ancestors; forth, stand in awe of "four knows" and do right things; fifth, stay away from five punishments and don't commit crimes in case of being killed; sixth, be open-minded if being official to educate people; seventh, be clear-handed, thrifty, steady and honest; eighth, often learn Confucian classics to warn self; ninth, cultivate morality according to the standards of Confucius being a gentleman; tenth, don't forget to do more good deeds.

释义：程家的后世子孙，要做到：一、不要不孝敬父母双亲；二、不要忘恩负义背地里诋毁老师和长辈；三、不要做盗贼使先祖蒙受耻辱；四、要敬畏"四知"做到为人行仁义之事；五、要远离五刑，不要干违法犯罪之事，以免遭到杀戮；六、为政做官要宽厚以教化百姓；七、自己要做到廉洁节俭敦厚诚实；八、要经常学习儒家经典以警诫自己；九、要用孔子做君子的标准来修养自己的道德。十、不要忘记要多做善事。

程文德　Cheng Wende

嘉靖年间，程钲次子文德榜眼及第，官居吏部左侍郎、翰林学士，掌管詹事府，专职辅助天子登基事。他捐资建议扩建宗祠，去信宗族云："某思万派同源，千枝同本。今日程氏数百人，皆我公范公裔也，自数百人规之，虽各有身；自公范公观之，实同一气，则奚可散而不属，疏而不亲也。是故先王制礼，尊祖而合族焉，尊祖于祠以亲幽也，合族于规以明亲也。幽明协和，祖孙各得，此之谓致一。

During Jiajing reign, the second son of Cheng Ji, Cheng Wende was the left Shilang of the Ministry of Personnel and scholar of the Imperial Academy. He took charge of Zhanshifu and exclusively assisted the prince to ascend the throne. He made a donation and suggested to expand the ancestral hall, writing a letter to the ancestral family, saying, "I thought that many schools all came from one origin. Today, several hundred people in our family were all the descendants of Gongfan. Everyone had his own living. Gongfan thought everyone should be bound to each other, be close to each other. Considering the rituals formulated by ancestral king, we should give respects to our ancestor and gather in one family. We should give sacrifices in the hall to show our peace, and follow the rule together to know our relatives. To achieve such harmony, everyone can have what he wanted.

我程氏家规行之已久，今某叨官于此，尚望伯叔兄弟敦行不怠。至于祠堂之充扩，昔欲通族共成，为人人欲尽其心也，即徐思之，今日公范公下，以受禄者惟某也，分禄以建祠，某不当独任耶？但官中俸薄口众，以书告兄弟以家租代禄，早完此工，不知已落成否？如其未也，某之责也，乞协心相事，兹寄去俸银若干，未足更助之岩石之租，尊祖于斯，合族于斯。凡我长幼皆翕然仁让，无少不亲不逊之事，则同气同心，祖宗虽远而若近；子孙虽多而若一，公范公知之，宁不欣然于地下耶？惟各相成，无穷之幸。"（程文恭公遗稿），自此，"闰三公祠"扩建成为"程氏宗祠"，为二进三开间及外建门楼，颇具规模的程氏总祠，也称程氏大宗祠。体现了程文德的孝悌之道。

Our family rules had been conducted for a long time. Today, I hoped anything wrong in what I said can be forgiven by our family members. As for the expansion of the ancestral hall, it should be conducted by everyone in our family. However, I thought I was the only official in this family, so I should give out my salary to shoulder the expansion construction. But my salary was little, so I wrote to my brother and told him to use my family rent, in order to make it finish quicker. Did it finish now? If it hadn't finished, I felt it my fault. I would try my best and would send some of my salary, in order to respect our ancestors and gather our family. All people in our family were easy-going and warm-hearted, never had offenses. We were of one mind, making our remote ancestor close to us. Although we had many offspring, they were united as one. If Gongfan knew this, he would be very happy. Helping each other will bring endless happiness." (remained manuscript of Cheng Wengong). Since then, "闰三公祠 (Run San Gong Hall)" was expanded to "Chengs' ancestral hall", with two rows and three rooms in each row. A gate tower was built outside the hall, making a big-scaled ancestral hall, also named Chengs' great ancestral hall. This showed the filial piety of Cheng Wende.

八、义乌　Yiwu

1. 石版塘宗氏　The Zongs of Shibantang

义乌宗氏始祖宗道溥（898—985，又名溥，字惟源）于五季之时，从河南省南阳避乱江南，卜居浙江省义乌沙溪石版塘。

The first ancestor of the Zongs of Yiwu was Zong Daopu (898—985, also named Pu, his courtesy name was Weiyuan) came to Shibantang, Shaxi, Yiwu, Zhejiang for living from Nanyang, Henan, in order to avoid the chaos of the Five Dynasties.

宗氏家庙全景 **Overall view of the Zongs' Family Temple**

家训摘选 **Selection of Family Instructions**

族中子弟当各勤生理：士者攻其学，农者力其耕，工者专其艺，商者蓄其货。毋学赌博以废事业，毋耽酒色以乱德性，毋摇唇鼓舌以生是非，毋游手好闲以荒岁月，毋狥利以害义，毋玩法以犯刑。凡此数端，各宜警省。

Family members should personally manage their own livelihood: students try their best to study, the farmers focus on their farming, the workers concentrate on their skills and businessmen save more products. Don't learn gambling and waste business, don't be addicted to wine and women and lose morality, don't make gossip and have quarrels, don't loaf around and waste life, don't pursue interest and sacrifice friendship, don't play with laws and commit crime. All members should be alerted to such things.

子孙不能守法妄作非为者，父母当严加教诲。不悛宜鸣家长会众戒责，更不悛或呈官究治，甚则削谱。怜孤恤寡王政所先，况骨肉之间非他人比也，族有此等须另加看顾，勿令失所。

If the offspring cannot obey the laws and do wrong, the parents should strictly teach them. If refusing to correct, they will be punished in public; if worse, they would be sent to the government, and even removed out of the genealogy. It's the duty of the court to take care of orphans and widows, not to mention the blood and bone ties with family members. If there were someone fallen into such difficulties he must be looked after specially and cannot become homeless.

宗泽　Zeng Ze

宗泽（1060—1128），北宋末、南宋初抗金名臣。字汝霖，义乌人。北宋嘉祐四年十二月十四日生，早年家境贫寒。元祐六年（1091）应进士试，对策陈时弊，考官恶其直言，抑为"同进士出身"录取。自此历任大名馆陶尉，浙江龙游、山东胶州及登州掖县县令，勤政爱民，治绩卓著，名声远扬，但得不到朝廷的赏识。

Zong Ze(1060—1128), was a famous official fighting with the Jin country during the late Northern Song Dynasty and early Southern Song Dynasty. His courtesy name was Rvlin. He was born on the 14th day of the 12th lunar month, the fourth year of Northern Song Jiayou reign in Yiwu. When he was young, his family was very poor. He took the imperial examination at the sixth reign of Yuanyou (1091). When he was answering the strategies to govern the country and demonstrating current malpractice, the examiner hated him speaking straightly, enrolling him "same as a successful candidate". Since then, he acted as the captain of Daming and Guantao, county magistrate of Longyou, Zhejiang, Jiaozhou and Ye county, Dengzhou, Shandong. He was hard-working and loved his people. Although with excellent achievements and good fame, he was not appreciated by the imperial court.

宣和元年（1119），因反对朝廷连结女真征契丹，被贬提举鸿庆宫，于是上表引退，拟在东阳山起兵，广集粮饷，防止敌人进拢。不久，升任河北义兵都总管，率军救真定。宗泽先以神臂弓挫敌凶焰，后纵兵进击，破金兵 30 余寨，斩敌数百，所获羊马金帛全部赏将士。此时康王赵构赴金议和至磁州，宗泽叩马劝止，乃留相州。

At the first year of the Xuanhe reign(1119), he was demoted to govern Hongqinggong because he went against the court that collaborated with Nvzhen to conduct a crusade against Qidan. Hence, he proposed to resign and planned military operation at Dongyang Mountain. He widely collected provisions and funds for troops in order to prevent enemies to come close. Later on, he was promoted to be the governor of Yibing, Hebei and led his army to save Zhending. Zong Ze firstly used Shenbi bows to crack down the arrogance of enemies, then ordered soldiers to attack. They destroyed over 30 camps of Jin soldiers and killed several hundred soldiers. The captured sheep, horses and golds and silks were all bestowed to soldiers. At the moment, King Kang, Zhao Gou arrived at Cizhou for a reconciliation with Jin country. Zong Ze dissuaded the King right away and kept him staying in Xiangzhou.

是年冬，宋钦宗任康王为兵马大元帅，宗泽为副帅。泽率军趋李固渡，途中遇敌，大破之。次年正月，率军至开德，与敌人打了 13 仗，仗仗获胜。建炎元年（1127）六月，宗泽以 67 岁高龄任东京留守，知开封府，招聚义兵近 200 万人，分署京郊 16 县，与金兵隔黄河对峙。此时岳飞投奔宗泽，泽见而奇其才，给以 500 骑兵，要其奋勇立功。岳飞听命而行，歼灭了敌人。从此岳飞就在宗泽部下南征北战。建炎二年正月，金人大举入侵，泽又大破之，金溃不成军，尽弃辎重。自此宗泽威震天下，金人畏惮宗泽，称之"宗爷爷"。

At the winter of this year, Song Emperor Qinzong appointed King Kang as generalissimo and Zong Ze as assistant general. Zong Ze led the army to Ligudu and met enemies midway and won the battle. In the first lunar month of the next year, he led the army to Kaide and had 13 battles with the enemy. And, they won all the battles. In 6th month of the first year of Jianyan reign(1127), Zong Ze of 67 years old stayed in Dongjing and governed Kaifeng Mansion. He gathered militiamen of nearly 2 million, who were deployed at 16 counties, confronting with Jin soldiers across the Yellow River. At the moment, Yue Fei went to Zong Ze's place for shelter. Zong Ze appreciated Yue's talents, so he gave Yue 500 cavalrymen and ordered him to bravely make achievements. Yue Fei followed the order and killed the enemy. Since then, under Zong Ze's command, Yue Fei fought with the enemy around the country. In the first lunar month of the second year of Jianyan reign, Jin people made a huge crusade, which was combated by Zong Ze. Jin army was completely routed and threw away supplies and gears for troops. Since then, Zong Ze's majesty was felt throughout the whole country, making Jin people very scared and named Zong Ze "Grandpa Zong".

建炎元年七月起，宗泽一年上疏 24 次，这就是著名的《乞回銮疏》，力劝宋高宗还京，以图恢复北方失地，均为奸佞所阻。宗泽忧愤成疾，疽发于背，明知自己病重，在世不长，却还是念念不忘请求赵构回銮开封，誓师北伐。

建炎二年七月癸巳日（1128 年 7 月 29 日）临终前，他对前来探望的将领沉痛地说："我以二帝蒙尘，悲愤至此，你们能多歼灭敌寇，那我死而无恨！"又不停地念诵杜甫名句："出师未捷身先死，长使英雄泪满襟。"闻者无不泣涕。直至断气，无一语及家事，惟连呼"渡河！渡河！渡河！"而逝。子颖与部将岳飞护柩至镇江，与夫人陈氏合葬于京岘山麓。后赠观文殿学士，通议大夫，赐谥忠简。

Since the 7th lunar month of the first year of Jianyan reign, Zong Ze had presented memorial to the throne for 24 times, namely famous *Qi Hui Luan Shu*. In order to regain the lost Northern territory, he tried his best to persuade Song Emperor Gaozong to return the capital. However, his requirements were all stopped by crafty sycophants. Zong Ze was so angry that he got sick and had subcutaneous ulcer on his back. He was well aware of his severe illness and didn't have much time, but he still constantly begged Zhao Gou to return Kaifeng and aim at northern expedition.

On July 29, 1128, before he died, in deep sorrow he said to the military leaders who came to visit him, "I felt ashamed of two Emperors. Full of grief and indignation, I hope that you can kill more enemies, then I have nothing to regret even if I died." Then he constantly recited the famous poems of Du Fu："Soldiers died before winning the war, it makes heroes cry all the time". It made all people present cry. Until he died, he never spoke of his family and only yelled, "Cross river! Cross river! Cross river!" Then he died. His son, Ying, and military officer Yue Fei escorted his coffin to Zhenjiang, and buried together with his wife Mrs Chen at Jingxian mountain. Later, he was conferred as the scholar of Guanwen hall, Tongyi official and posthumous title as Zhongjian.

2. 吴氏　The Wus

据各宗谱所载，吴氏迁居主要有四支，两支迁于北宋，两支迁于清代。其一，大元吴氏。北宋治平年间（1064—1067），苏州吴县人吴造以父荫补乌伤主簿，卒葬邑东十五里石板岭（大元村附近），其幼子吴德，遂徙居园圃（即大元村），吴造为义乌吴氏始迁祖。其二，前洪吴氏。北宋末年，为避金兵入侵，郡马吴商霖随宋室南迁，其子吴琥任吴宁县令，致仕后定居苏溪八里，吴琥第四子吴权，于南宋乾道六年（1170），迁居莲台（北苑街道前洪村）。其三，清咸丰年间（1851—1861），稠江街道象山村吴氏始迁祖由东阳白坦村迁入。其四，清嘉庆元年（1796），佛堂镇稽亭村吴氏始迁祖吴梧孙从福建莆田迁入。

According to all genealogies, there were mainly four branches of the Wus settling down. Two branches moved during the Northern Song Dynasty, and the rest two moved during the Qing Dynasty. The first was the Wus of Dayuan. During the Zhiping reign of the Northern Song Dynasty, Wu Zao, a man in Wu County of Suzhou obtained the deputy governor of Wushang due to the official position of his father. After death he was buried at Shibanling, 15 li away from the east of the county(near Dayuan village). His young son moved to Yuanfu(namely Dayuan village). Wu Zao was the first ancestor of Yiwu Wus that moved here. The second was the Wus of Qianhong. At the late Northern Song Dynasty, in order to dodge from Jin soldiers' invasion, Wu Shanglin, the husband of the princess, moved south along with the Song imperial family. His son, Wu Hu acted as Wuning County's magistrate. After retiring, he settled at Bali, Suxi. His fourth son, Wu Quan moved to Liantai(Qianhong village, Beiyuan street) at the sixth year of the Southern Song Qiandao reign(1170). The third was the first ancestor of Wus in Xiangshan village, Choujiang Street, who moved here from Baitan village, Dongyang at the Qing Xianfeng reign(1851—1861). The fourth was Wu Wusun the first ancestor of Wus in Jiting village, Fotang town, who moved here from Putian, Fujian, at the first year of the Qing Jiajing reign(1796).

吴氏家训（摘选）The Wus' Family Instructions（excerpts）

子孙有习嫖赌者会众呈公责治；犯奸盗者呈官外更削谱名，至习优者规责不听亦削谱名，若其子改行仍谱其子。

If there were offspring having dissipated life, they would be sanctioned in public; bad people and thieves would be kicked out of genealogy, people who don't listen to expostulation would be kicked out of genealogy. If his son changed behaviors, the name of his son would be written on the genealogy.

兴家在礼义，礼义由读书。读书入泮者，本祠优给银一两，科举给银一两，中乡榜给银陆两，甲榜给银十二两。明经给银三两，太学给银一两，文武乡榜甲榜给银视文榜之半杂流，文武职员给银一两，至出仕后各视其力蠲银入

大元村七幢 **No. 7 Building of Dayuan village**

祠，倍酬优给之数，其肯照例加厚者更见孝慈诚心。

The key to prosper a family was rite and morality, which can be achieved by studying. When a child went to school, the hall would give him one liang of silver, and one liang of silver for participating in imperial examination, six liang of silver if passing the exam, and twelve liang of silver if being the top candidate. Three liang of silver for senior licentiate, one liang of silver for being a student in imperial college, half liang for top in civil and military county examination, one liang of silver for civil official and military officer. After being an official, they would donate money to the hall according to their conditions, and more money stands for more sincere filial piety.

吴百朋　**Wu Baipeng**

吴百朋（1519—1578），字惟锡，号尧山，浙江义乌人。明嘉靖二十六年进士。官至刑部尚书，与戚继光生活于同一时代，为抗倭寇、平内乱、固边防立下了不朽功勋，系一代名闻遐迩的儒将。

Wu Baipeng (1519—1578), styled himself as Weixi and titled himself as Yaoshan. He was born in Yiwu, Zhejiang. He was a successful candidate in imperial examination at the 26th year of the reign of Ming Emperor Jiajing. He acted as Minister of the Ministry of Punishments. He and Qi Jiguang were contemporary. He made immortal achievements in fighting against Japanese pirates, pacifying civil strife and consolidating border defence, reputed as a famous general who was an equally accomplished man of letters.

吴百朋家境贫寒，自幼丧母，饱受生活的磨难。但他穷且弥坚，勤学好问，立志报效国家，为黎民百姓做事。嘉靖二十六年（1547），吴百朋29岁，赴京参加殿试，金榜题名，登三甲进士。吴百朋为官数十年，清正廉洁，一生俭朴，所居房屋甚是简陋，只能挡风避雨。衣服、被褥，均属普通，并不华丽。有一年寒冬，吴百朋巡视雕鹗堡，家里人给他做了件绣花的新棉袍，吴百朋坚持不肯穿。将士们见了，深受教育，都主动脱去身上的锦衣绣袍。

Born in a poor family, Wu Baipeng lost his mother when he was young and suffered from the hardship of the life. Although he was poor, he was tough, diligent and inquisitive. He aimed to serve the country and work for common people. At the 26th year of the Jiajing reign(1547), Wu Baipeng of 29 years old, went to the capital to participate final imperial examination. He passed the examination and was top three candidates. Being an official for over ten years, Wu Baipeng was upright and clear-handed. He lived a thrifty life with a rough house which could only keep out the wind and the rain. His clothes and bedding were all normal. In a winter of one year, when Wu Baipeng made an inspection tour to Diao'e Fortress, his family made him a new embroidered cotton wadded robe, but Wu Baipeng refused to wear. His soldiers saw this and were deely touched, so they took off their beautifully embroidered robes.

吴百朋巡抚虔州，抗倭平乱六年。离任时，将他按例应得的"逾额者十七万金"悉数上交国库。离任时单车就道，一无所携，其清廉之德深得部属的敬佩。吴百朋一生勤奋好学，其主要的传世之作有《吴百朋奏疏》3卷，《南赣督抚奏议》存一、三、七卷（余佚失）及部分诗稿等。其他著作如《抚虔志》《用兵纪实》等惜已佚失。

Wu Baipeng governed Qianzhou, and spent six years in fighting against Japanese pirates and pacifying civil wars. When he quit his job, he handed in his deserved "170, 000 of ratio excess" to the exchequer. He rode a single car with nothing, whose clear-handed morality greatly won the respect of his subordinates. Wu Baipeng was diligent and studious in his all life, whose main existing works are three volumes of *Wu Baipeng's Memorial to the Throne*, the 1st, 3rd, and 7th volume of *Nan Gan Du Fu Zou Yi* (other volumes lost) and some poems. Other works like, *Governing Qian* and *War On-the-spot Report* were lost.

吴百朋从严治家，倡导耕读家风。他常常教育子孙要自立自强，好自为之。其夫人王氏因相夫教子，贤德闻名乡里，被朝廷封为诰命一品夫人。常言道："君子之泽，五世而斩。"而吴百朋家族却不尽然，历代均有人才涌现。

Wu Baipeng ruled his family strictly and advocated part-work and part-study family tradition. He often educated his offspring to be self-dependent and strong and conduct themselves well. His wife, Mrs Wang assisted her husband and brought up children. Her virtuous morality was famous in the hometown, titled with First-class Madam by the court. As it is often said, "the conduct and morality of a gentleman will vanish after several generations' inheritance." However, it was not true for Wus' family for there were so many talents in his family in every generation.

子吴大瓒，官至知府同知，著有《抱膝居稿》；孙吴存中，擅书法，著《书论》2 卷、《字学》10 卷；曾孙吴之器，举人，官至兵部车驾司主事，著有《婺书》《婺书别录》《明月斋稿》等；曾孙吴之艺，举人，著有《得山堂文稿》12 卷、《诗稿》26 卷，等；曾孙媳倪仁吉，乃一代才女，工书画、精音律、善刺绣，著有《凝香阁诗稿》。

His son, Wu Dazan, acted as a magistrate of district and wrote out *Bao Xi Ju Gao*. His grandson, Wu Cunzhong was good at calligraphy and wrote 2 volumes of *Shu Lun* and ten volumes of *Zi Xue*. His great-grandson, Wu Zhiqi, was a successful candidate in the county examination and acted as the chief of imperial chariots of the Ministry of War. He had written *Book of Wu*, *Wu Shu Bie Lu*, *Ming Yue Zhai Gao*. His great-grandson Wu Zhiyi, was also a successful candidate, and had 12 volumes of *De Shan Tang Wen Gao* and 26 volumes of *Verse Manuscripts*. His great-granddaughter-in-law, Ni Renji, was a talented woman, adept at calligraphy, painting, instruments and stitchwork. She had written *Ning Xiang Ge Verse Manuscripts*.

第三部分 赏析·家规
Appreciation · Family Rules

《郑氏规范》 *Zhengs' Family Regulations*

第一条 立祠堂一所，以奉先世神主，出入必告。正至朔望必参，俗节必荐时物。四时祭祀，其仪式并遵《文公家礼》。然各用仲月望日行事，事毕更行会拜之礼。

Build an ancestral temple to worship the ancestral shrine. Every important affair should be reported to the ancestral temple. On the first day and fifteenth day of each lunar month, every family member must participate the worship ceremony in the ancestral hall. They serve fresh fruits to their ancestors during traditional festivals. The sacrificial ceremony in four seasons should comply with *Family Rites of Wen Gong* (Zhu Xi). The dates for sacrificial ceremony are on the fifteenth day of the second, the fifth, the eighth and the eleventh lunar month. After the sacrificial ceremony, descendants will worship on bended knees.

释义：建祠堂一所，以供奉先祖神位，家族有重大事务必到祠堂禀告祖先。到每月初一、十五日必须在祠堂举行参拜仪式，逢传统节日必须敬奉时鲜果品。春夏秋冬四时祭祀仪式都应遵照《文公家礼》。但举行祭祀仪式的日子为每季季中月份的十五日，即夏历二月十五日、五月十五日、八月十五日、十一月十五日。祭祀完毕，再举行会拜之礼。

第二条 时祭之外，不得妄祀徼福。凡遇忌辰，孝子当用素衣致祭。不作佛事，象钱寓马亦并绝之。是日不得饮酒、食肉、听乐，夜则出宿于外。

Besides the four-season sacrifice, family members cannot violate the provisions to sacrifice for blessing. Whenever there is the death anniversary for elders, the filial son should wear a plain dress and hold a sacrificial ceremony. But it is not allowed to use the paper money, the paper men and paper horses. On that day, the filial son mustn't drink alcohol, eat meat or watch the opera. He should wait outside the ancestral temple at night.

释义：除四时祭祀之外，不得随意违反规定祭祀求福。凡遇长辈忌辰，孝子应该穿着素衣，举行祭礼。但不作佛事，并不得使用冥钱、纸人纸马。当日孝子不得饮酒、食肉、听戏，晚上必须宿在祠堂外面守候。

第三条 祠堂所以报本，宗子当严洒扫扃钥之事，所有祭器服不许他用。祭器服，如深衣、席褥、盘盏、碗碟、椅桌、盥盆之类。

The ancestral hall is the place to remember the ancestor's kindness and trace the source. The eldest son must strictly manage sanitary duties, such as watering and sweeping, and security affairs such as closing and locking the door. All sacrificial vessels and gowns are not allowed to put into other uses, such as *shenyi* (formal dress), mats, dishes, tables and basins and so on.

释义：祠堂是怀念祖先恩德、溯本追源的场所。宗子必须严格管理洒水扫地等保洁以及关门上锁等安全事务，所有祭器礼服不得移作他用。祭器礼服包括深衣、席褥、盆盏、碗碟、椅桌、盥盆之类。

第四条 祭礼务在孝敬，以尽报本之诚。其或行礼不恭，离席自便，与夫跛倚、欠伸、哕噫、嚏咳，一切失容之事，督过议罚。督过不言，众则罚之。

While performing the sacrificial ceremony, people should show their respect and loyalty to their ancestors. If someone displays any disrespect during the ceremony, such as leaving the seat randomly, standing in an ease pose, yawning, stretching one's arms, hiccuping, sneezing and coughing. All of these improper behaviors will be punished by the Supervisor. If the Supervisor didn't put forward any punishment, then everyone at present would discuss and come up with some punishments.

释义：祭祀务必孝敬，以向先祖表达报答恩德的诚心。如果有人在行礼之时不恭敬，随便离开席位，站立不正、打哈欠、伸懒腰、打呃、打喷嚏、咳嗽，一切失容之事，由督过出面提出处罚。如督过不言，大家议罚。

第五条 拨常稔之田一百五十亩，世远逐增，别蓄其租，专充祭祀之费。其田寿印"义门郑氏祭田"六字，字号步亩亦当勒石祠堂之左，俾子孙永远保守。有言质鬻者以不孝论。

150 mu (10 hectares) of fertile fields would be allocated specifically for the sacrificial use, and the areas of the fields will increase as the time goes by. The landbook is printed with "义门郑氏祭田" and the location and the acres of the sacrificial field are inscribed on the stone tablet, which stands on the left side of the ancestral temple. The descendants will safeguard the sacrificial fields in case someone would put them for sale or mortgage. Those who propose to sell or mortgage them are unfilial.

释义：拨常熟良田一百五十亩，而且今后逐步增加，将其租谷另行蓄藏，专用于祭祀所需的费用。其祭田券契印上"义门郑氏祭田"六字，并将祭田坐落、亩分刻于石碑，立于祠堂的门首左面，使子孙永远保卫坚守。如有人提议将祭田抵押、出卖的，作不孝论。

第六条 子孙入祠堂者，当正衣冠，即如祖考在上，不得嬉笑、对语、疾步。晨昏皆当致恭而退。

When the descendants enter the ancestral hall, they should be properly dressed, just as the ancestors themselves were there, not to laugh or talk, to walk slowly and steadily, not to walk quickly, and to enter and leave the ancestral hall with great respect in the morning and dusk.

释义：子孙进入祠堂，应当衣冠端正，犹如先祖亲自在上，不得嬉笑、谈话，走路要稳、慢，不得快步，早晨和黄昏进入祠堂时都应该极其恭敬地进退。

第七条 宗子上奉祖考，下壹宗族。家长当竭力教养，若其不肖，当遵横渠张子之说，择次贤者易之。

The eldest son in the family should shoulder the responsibilities to serve the ancestors and unite the whole family. Parents must strive to educate him to meet the requirements of the firstborn son. If he is incompetent, parents should follow what Zhang Zai, a Neo-Confucianist in the Northern Song Dynasty, has ever suggested, that is, to choose another capable child to replace him.

释义：家族中的嫡长子作为宗子上要事奉祖先，下要凝聚全族。家长必须竭力教育他使其合乎宗子的要求，若其不称职的，应当遵照北宋理学家张载之说，另选择贤能的人担任。

第八条 诸处茔冢，岁节及寒食、十月朔，子孙须亲展省，妇人不与。近茔竹树不许剪拜，各处庵宇更当葺治。至于作冢制度，已有《家礼》可法，不必过奢。

During the Lunar New Year, Qing Ming and the first day of the 10th month, the descendants must go to visit the cemetery, but women are not allowed to go with them. Bamboo forests and trees near the cemetery are not allowed to be cut down. All temples should be repaired in good time. The standards for constructing the mausoleum are to be in accordance with *Family Rites*, not to be too extravagant.

释义：各处祖先坟茔，每逢年节、清明及十月初一，子弟须亲自前往墓地检查察看，但妇女不得一同前往。墓地附近的竹林、树木不许砍伐。各处庵堂、庙宇应当及时修葺。至于修筑陵墓的标准，可遵照《家礼》，不必过于奢侈。

第九条 坟茔年远，其有平塌浅露者，宗子当择洁土益之，更立石深刻名氏，勿致湮灭难考。

If the tomb is in disrepair for a long time, in case the coffin sunk, collapsed or exposed because it is buried too shallow to reveal the coffin, the eldest son should choose clean soil to be added to the graveyard and then put up a tombstone and carve the name of the death in case the name is difficult to verify.

释义：坟茔年久失修，若有低陷、坍塌以及掩埋过浅露出棺木的，宗子应当选择洁净的泥土添加在坟丘上，再立上墓碑，刻明逝者姓名，以免逝者湮灭，难以考证。

第十条 四月一日，系初迁之祖遂阳府君降生之朝，宗子当奉神主于有序堂，集家众行一献礼，复击鼓一十五声，令子弟一人朗诵谱图一过，曰明谱会。圆揖而退。

The first day of the 4th month is the birthday of the first ancestor, Zheng Huai, who first moved in Pujiang. Zongzi should worship the ancestor Zheng Huai and gather all the families to present salute in the Hall of Order. And then he would beat the drum fifteen times and ask a child to recite Zhengs' ancestors, called "Knowing the Ancestors Gathering". After the recite, all families would salute and retreat in order.

释义：四月初一日，是初迁先祖淮公诞生吉日。宗子应当奉淮公的神主牌位置有序堂上，会集家众行一献礼，然后击鼓十五声，令一子弟朗诵郑氏先祖世系一遍，称作"明谱会"。朗诵完毕，大家互相行揖礼后按次序退下。

第十一条　朔望，家长率众参谒祠堂毕，出坐堂上，男女分立堂下，击鼓二十四声，令子弟一人唱云："听，听，听，凡为子者必孝其亲，为妻者必敬其夫，为兄者必爱其弟，为弟者必恭其兄。听，听，听，毋徇私以妨大义，毋怠惰以荒厥事，毋纵奢侈以干天刑，毋用妇言以间和气，毋为横非以扰门庭，毋耽曲蘖以乱厥性。有一于此，既殒尔德，复隳尔胤。眷兹祖训，实系废兴。言之再三，尔宜深戒。听，听，听。"众皆一揖，分东西行而坐。复令子弟敬诵孝悌故实一过，会揖而退。

On the first and the 15th day of each month, the patriarch led their families to visit the ancestral hall. After that, they came out of the hall and were separated into two groups: male and female. Then one child beat the drum 24 times and chanted: "Listen! Listen! Listen! Every son must be filial to his parents. A wife must respect her husband. A brother must love his younger brother and the younger brother must be polite to his elder brother. Listen! Listen! Listen! Do not harm the righteous cause because of private affairs. Do not get slack at your business because of your laziness. Do not indulge and be punished by the heaven. Do not harm the harmony of the big family because of your wife's slanderous talk. Do not do evil to degrade the family. Do not indulge in wine and food and forget a man's real nature. If there are any of these actions, it will damage your own moral standing and destroy your children. Keep in mind this ancestral precept, it is related to the rise and fall of a family! Repeatedly and repeatedly I tell you these and you should rethink them deeply. Listen! Listen! Listen!" After listening, people saluted and sat face-to-face. Then they listened to some report about the filial stories of their ancestors. After that, they saluted and retreated.

释义：每月初一、十五日，家长率族众参谒祠堂完毕，出坐于堂上，男女分立堂下。然后击鼓二十四声，由一名子弟高声吟唱训词："听！听！听！凡是做儿子一定要孝顺父母，做妻子的一定要尊重丈夫，做兄长的一定要怜爱弟弟，做弟弟的一定要礼待兄长。听！听！听！不要为了私情而妨害大义，不要因为懒怠而荒废正事，不要纵情奢侈而受到老天的惩罚，不要因为听信内人的谗言而伤了家庭和气，不要横行作歹而扰乱门庭，不要沉溺于酒食而损害了人的本性。如果有以上一种行为，就一定会损害你自己的道德名望，又会毁灭你的子孙后代。牢记这个祖训，实在是关系到一个家庭的兴衰！反复再三向大家说明，你们应该深刻地引以为戒。听！听！听！"众人听后皆作揖，分东西向而坐。又令子弟敬诵已故先祖孝悌的事迹后，相互拜揖后退下。

第十二条 每旦，击钟二十四声，家众俱兴。四声咸盥漱，八声入有序堂。家长中坐，男女分坐左右，令未冠子弟朗诵男女训诫之辞。《男训》云："人家盛衰，皆系乎积善与积恶而已。何谓积善？居家则孝悌，处事则仁恕，凡所以济人者皆是也；何谓积恶？恃己之势以自强，克人之财以自富，凡所以欺心者皆是也。是故能爱子孙者遗之善，不爱子孙者遗之恶。《传》曰：'积善之家必有余庆，积不善之家必有余殃。'天理昭然，各宜深省。"《女训》云："家之和与不和，皆系妇人之贤否。何谓贤？事舅姑以孝顺，奉丈夫以恭敬，待娣姒以温和，接子孙以慈爱，如此之类是也；何谓不贤？淫狠妒忌，恃强凌弱，摇鼓是非，纵意徇私，如此之类是也。天道甚近，福善祸淫，为妇人者不可不畏。"诵毕，男女起，向家长一揖，复分左右行，会揖而退。九声，男会膳於同心堂，女会膳於安贞堂。三时并同。其不至者，家长规之。

Every morning, after the 24 beats of the bell, all family members should get up from the bed. After the 4 beats, they should finish the washing and combing. After the 8 beats, they should enter the Hall of Order. The patriarch sat in the front and the male sat on the left side and the female on the right side. A child who is under the age of 16 will recite the Principles of men and women. The content of *Male Principles* is: "The rise and fall of a family is closely related to the accumulation of good or evil. What is good? At home, we should perform filial piety. Benevolence and forgiveness should be considered when dealing with things. It is also a good deed to help others. What is evil? It is evil to force others, to bully others, to deduct other people's money and enrich themselves. Those who want to bully others are evil people. Therefore, those who love their children and grand children will teach their children to be good, and those who do not love their children will indulge their children and let them do evil. A book called *Yi Jing* (The Book of Changes) said: 'A family with accumulation of good will receive extra luck and the family with accumulation of evil will eventually receive bad luck'." The content of *Female Principles* is: "The harmony and disharmony of a family have something to do with whether the women in the family are virtuous or not. What is virtuousness? Serve her father-in-law and mother-in-law filially, treat her husband respectfully, deal with other family members tenderly and love her children and grandchildren. Such things are called virtuousness. What is viciousness? If her attitude is not solemn, her posture is frivolous, if she is jealous of others or bullies others, if she likes to stir things up or if she is too selfish, she is vicious. Heaven's retribution is very fast. Let those who do good be blessed and those who do debauchery suffer. Women must bear a heart of reverence." After reciting, all family members stood up and bowed to the patriarch, and then they saluted each other and retreated. With nine bells ringing, the men went to the Concentric Hall for dinner and the women went to the Anzhen Hall for dinner. All three meals were the same. Those who were not present would be punished according to the family rules.

释义：每天早上击钟二十四声，家众都应起床。击钟四声梳洗完毕，八声入有序堂。家长坐在堂中，男女家众分坐左右，令一名未及十六岁子弟朗诵男女训诫。《男训》内容是："一户人家的盛衰，都与积善还是积恶有密切关系。什么是积善？在家则孝悌为先，处事要仁爱宽恕，凡遇事能接济他人等都是善事。什么是积恶？恃自己之势强迫欺侮他人，克扣别人的钱财而充实自己，凡有欺心存在的皆是。所以爱子孙的人家就教育子弟从善，不爱子孙的就放纵子弟任其行恶。《易经传》云：'积善之家，必有余庆，积不善之家，必有余殃。'天理昭然，大家都应该认真思考。"《女训》的内容是："一个家庭的和睦与不和睦，与家庭的妇女是否贤惠有关。什么是贤惠？侍候公婆以孝顺，侍奉丈夫以恭敬，对待妯娌以温和，接待子孙以慈爱，如此之类就叫作贤惠。什么是不贤惠？态度不庄重，体态轻浮，妒忌他人，恃强凌弱，拨弄是非，纵意徇私，如此之类就叫作不贤。上天的报应是非常快的，让行善之人得福而让淫荡之人遭祸，作为妇人不可没有畏惧之心。"朗诵完毕，男女家众起立，向家长行一揖礼，分左右两行一起互相行揖礼，然后退下。鸣钟九声，男子到同心堂会膳，妇女到安贞堂会膳，三餐都相同。如家众有不到的，家长依据家规处理。

第十三条　家长总治一家大小之务，凡事令子弟分掌，然须谨守礼法以制其下。其下有事，亦须咨禀而后行，不得私假，不得私与。

The patriarch is in charge of all family affairs. All family affairs can be assigned to certain family members who are cautious and upright, and could be able to manage other family members according to family rules. All family affairs should be reported to the patriarch. Children could not make decisions on their own under the pretext of the patriarch.

释义：家长总治全家大小事务，所有事务可派子弟分别掌管，但是必须谨慎遵守礼法，以家范规定约束家众。家众有事，亦必须咨询禀告家长后才能执行，不得私自假冒家长名义，不得私自决定主张。

第十四条 家长专以至公无私为本，不得徇偏。如其有失，举家随而谏之。然必起敬起孝，毋妨和气。若其不能任事，次者佐之。

The patriarch should hold unselfishness and unbiasedness as the fundamental rule. If he had been at fault, the whole family could remonstrate with him. But exhortation method must take the filial piety as the criterion, in which the family harmony will not be harmed. If the patriarch is really not competent to manage the affairs himself, the one who is in the second place should assist him in his work.

释义：家长以至公无私为根本，不得以私情而有所偏向。如其有过失，全家随时可以规劝他。但规劝方式必须以孝敬为准则，不能伤害家庭和气。家长本人确实不能胜任管理事务的，由年龄居第二者辅佐他办事。

第十五条 为家长者当以诚待下，一言不可妄发，一行不可妄为，庶合古人以身教之之意。临事之际，毋察察而明，毋昧昧而昏，须以量容人，常视一家如一身可也。

As the patriarch, he should treat all family members honestly. He should not speak and act casually. It is hoped that the patriarch can act in line with the ancients' intention to lead as an example. When it comes to affairs, do not take too much care about the details to show shrewdness, and do not be confused about things. Be generous while making decisions and accommodate other people. Always love and take care of your family just like your own body.

释义：身为家长的，应当至诚对待家众，讲话不可随便，行动不可妄为，希望家长行事能够符合古人的以身作则之意。临事之际，不要在细节问题方面过于计较以显示精明，也不要糊涂待事，在决断时要大度，以量容人，平常爱护家庭如爱护自己的身体一样。

第十六条 家中产业文券, 既印"义门公堂产业子孙永守"等字, 仍书字号。置立《砧基簿》, 书告官印押, 续置当如此法。家长会众封藏, 不可擅开。不论长幼, 有敢言质鬻者, 以不孝论。

Family property vouchers are printed with "Yi Men Gong Tang property, descendants keep it forever" and some code names. The family board sets up an *Anvil Book* to record the vouchers, and the proofs of the vouchers will be sent to the Government in writing, and the subsequent purchase of the property will be handled alike. The patriarch will seal them in front of all family members and mustn't open them privately. Whether young or old, those who dare to say selling or mortgaging the property are unfilial.

释义: 家中产业券契, 都要印上"义门公堂产业子孙永守"等字样, 还需编字号。公堂设立《砧基簿》记录券契, 并将券契的印样花押书面禀告官府, 以后置办的产业亦如此办理。家长会集家众将其封藏, 不得擅自开启。族人不论长幼, 有敢言抵押出卖的, 以不孝论处。

第十七条 子孙倘有私置田业、私积货泉, 事迹显然彰著, 众得言之家长, 家长率众告于祠堂, 击鼓声罪而榜于壁。更邀其所与亲朋, 告语之。所私即便拘纳公堂。有不服者, 告官以不孝论。其有立心无私、积劳于家者, 优礼遇之, 更于《劝惩簿》上明记其绩, 以示于后。

If the children or grandchildren buy their fields privately and accumulate wealth with sufficient evidence, people should report to the patriarch. The patriarch should tell the public in the ancestral hall and beat the drum to show their sin. And their bad deeds are written on a piece of paper which is to be pasted on the wall. The patriarch would also invite other relatives and friends to exhort the privately-bought fields and money to be paid to the court. If the children or grandchildren do not accept the exhortation, they would be sued and labeled as unfilial descendants. For those who are selfless and industrious and those who do contributions to the family, the court will give them generous gifts and write their good deeds on *Book of Persuasion and Punishment* to show the descendants.

释义: 子孙如果私置田业、私积钱财, 事实非常明显的, 众人应禀告家长, 家长率众告于祠堂, 击鼓明示其罪, 并将劣迹书于纸上贴于墙壁。更邀其亲朋好友规劝将私置的田业、钱财交纳公堂。有不服者, 告官以不孝论处。对立心无私、积极劳动有利于家的人，公堂给以优厚的礼物，并于《劝惩薄》上写明其事迹，以示后人。

第十八条 子孙赌博无赖及一应违于礼法之事，家长度其不可容，会众罚拜以愧之。但长一年者，受三十拜；又不悛，则会众痛箠之；又不悛，则陈于官而放绝之。仍告于祠堂，于宗图上削其名，三年能改者复之。

If the descendants do gambling or other violation of the law, the patriarch should punish them in the face of public according to the severity of bad deeds. They will be made to bow to the elders and feel shameful. They will give 30 bows to those who are one year older than them. Those who do not repent will be flogged in public. If they still do not repent, parents will report to the government and put them into prison, and expel them from the family. And then erase their names from the genealogy in the ancestral hall. If they can repent within three years, their names will be rewritten on the genealogy.

释义：子孙有赌博无赖等一切违于礼法之事，家长根据事情发生的程度，给以酌情处理。其不可容的，当众罚其跪拜使其羞愧，每年长一岁受三十拜。不悔改的，则当家众面用鞭痛打。再不改者，则禀告官府将其逐出家门。然后在祠堂里当众从宗谱上削去名字。三年内能改的，将其名字复上谱图。

第十九条 凡遇凶荒事故，或有阙支，家长预为区划，不使匮乏。

In the event of unfortunate death or disaster, or the family income is not enough to support the family, the patriarch should plan for their children in advance, and do not let them sink into poverty.

释义：凡遇不幸死亡、灾难事故，或者收入不够开支的家众，家长要预先为其谋划，不使他们因此贫困。

第二十条 朔望二日，家长检点一应大小之务。有不笃行者议罚；诸簿籍过日不结算及失时不具呈者，亦量情议罚。

On the 1st and the 15th day of every month, the patriarch should check and count all family affairs, be they big or small. And penalties should be imposed on those who did not conscientiously perform their duties. Those who did not settle the book during that day and those who did not present the book on time will also be punished according to the situation.

释义：初一、十五二日，家长检查清点全家一切大小事务，对有不认真履行职责者要提出责罚，各种账册过日不结算的，及过时不呈报的，亦量情轻重议罚。

第二十一条　内外屋宇、大小修造工役，家长常加检点。委人用工，毋致损坏。

The patriarch should often inspect the building staffs of the inside and outside buildings. Since the matters have already been done by others, he should make sure that the property would not be damaged.

释义：对内外屋宇、大小修造的工作人员，家长经常要加以检查清点，既然把事情交给别人去办，再不可致使财物损坏。

第二十二条　每岁掌事子弟交代，先须谒祠堂，书祝致告，次拜家长，然后领事。

Every year, when the descendants in charge want to handle the handover, they must first visit the ancestral hall, and then write a message to the ancestors, after that, they will meet the patriarch to accept the new business.

释义：每年掌事子弟要办移交事务时，必须先参拜祠堂，书写祝辞向先祖禀告，然后拜见家长，接受新的事务。

第二十三条　设典事二人，以助家长行事。必选刚正公明、材堪治家、为众人之表率者为之，并不论长幼、不限年月。凡一家大小之务，无不预焉。每夜须了诸事，方许就寝。违者，家长议罚。

The family should arrange two descendants to a position called "Dian Shi" (steward) to help deal with the family affairs. They should select those who are righteous and capable to manage the whole family regardless of age as long as they behave themselves as the role models. Family affairs, be they big or small, should be prepared in advance. And "Dian Shi" should arrange the family affairs appropriately before going to bed every day. If they cannot perform their duties, they will be punished by the patriarch after discussion.

释义：设立典事一职由二人担任，以帮助家长处理日常事务。必须挑选刚正公明，能治家、能为众人做表率的人才，不论长幼，不限年月。凡一家大小事务，典事都应参加处理，每日晚上必须将一日之事料理清楚，方可就寝。违者家长议罚。

第二十四条　每夜会聚之际，典事对众商榷，何日可行某事，书之于籍。

上半月所书，下半月行之；下半月所书，次上半月行之，庶无迁滞之患。事当即行者弗拘。

Every night during the family gathering, "Dian Shi" will make an open discussion as to when something will be done. They record it in the book. What is recorded during the first half of the month will be done during the second half of the month; What is recorded during the second half of the month will be done during the first half of the next month. This could save the things from being delayed and stopped. The matters which should be done immediately do not belong to this type.

释义：每晚会聚之时，典事要当着家族众人公开商量确定何日行某事，并记载在书簿上，上半月所记，下半月完成，下半月所记的，下月上半月完成，不能拖延和停止不办。必须马上完成的事务时限不限。

第二十五条　择端严公明、可以服众者一人，监视诸事。四十以上方可，然必二年一轮。有善公言之，有不善亦公言之。如或知而不言，与言而非实，众告祠堂，鸣鼓声罪，而易置之。

The family should select one person who is impartial, upright and who is able to serve the public to monitor the family affairs. The person in charge of the surveillance must be at least 40 years old and hold the position for every 2 years. He should raise good or bad things in court. If he did not tell the truth or if he did not tell all the things he knew, people can sue him in the ancestral hall by beating the drums. Then the supervisor will be replaced by others.

释义：选择为人端正严明、并能服众的一人，监视家族各类事务。担任监视的人必须年满四十岁，且二年一任。家中有好事及不好之事，都由监视在公堂上提出。如知道后不提议的，或说得不切实际，家众可告于祠堂，鸣鼓声罪。然后更换并选择新的监视。

第二十六条　监视莅事，告祠堂毕，集家众于有序堂，先拜尊长四拜，次受卑幼四拜，然后鸣鼓，细说家规，使肃听之。

If the supervisor has something to tell the ancestral hall, he should gather the family members in the hall and firstly bow 4 times to the seniors, then the younger generations will bow to him 4 times. Then after the beats of drum, he will tell the family rules in detail, everyone present should listen carefully.

释义：监视遇到有事之时，告祠堂后，会集家众到有序堂，先拜尊长四拜，其次受其他人四拜，然后鸣鼓细说家规，家众必须肃穆静听。

第二十七条　监视纠正一家之是非，所以为齐家之则，而家之盛衰系焉，不可顾忌不言。在上者，必当犯颜直谏，谏若不从，悦则复谏；在下者则教以人伦大义，不从则责，又不从则挞。

The supervisor is responsible for correcting the wrong behaviors of a family, therefore he is vital in family management and closely related to a family's rise and fall. He should do his duties devotedly, in which he must advise frankly to both his elders and his younger generations. For the elders, he should speak his mind directly regardless of the "face". If the elders did not listen, he should say it again and again when they are in good moods. For the younger generations, he should teach them ethics and morality and how to behave as a good man. If they did not listen, he should rebuke them and even flog them.

释义：监视及时纠正整个家族的是非，是实现治家原则的保证，家道的兴衰全部维系在监视的身上，为此监视对家族事务不可顾忌不言。对于尊长必须不顾脸色直言规劝，若尊长不听，等到他高兴的时候再规劝。对年幼下辈，教以人伦大义，怎样做人，如不听则加以斥责，又不从则用鞭打。

第二十八条　立《劝惩簿》，令监视掌之，月书功过，以为善善恶恶之戒。有沮之者，以不孝论。

The family should set up a *Book of Persuasion and Punishment*, which is mastered by the supervisor. Every month, the supervisor should record the merits and bad deeds in the book as an encouragement and a warning. Those who want to stop it are unfilial.

释义：设立《劝惩簿》一本，由监视掌握，每月将家众的功与过记载在《劝惩簿》上，作为警示，有阻止者以不孝论。

第二十九条　造二牌，一刻"劝"字，一刻"惩"字，下空一截，用纸写贴。何人有功，何人有过，既上《劝惩簿》，更上牌中，挂会揖处，三日方收，以示赏罚。

There are two wooden plates, one carved "persuasion" and the other "punishment". Under the inscription, there is some empty place to paste posters with someone's merits and bad deeds on it. Good deeds and bad deeds will not only be put on the *Book of Persuasion and Punishment*, but also on the wooden plates for 3 days in the saluting hall to show praise and punishment.

释义：制造木牌两块，一块刻"劝"字，一块刻"惩"字，下半部空出一段。何人有功劳，何人有过失，不仅记入《劝惩簿》，还要写在纸上分别贴到"劝""惩"二牌的下半截空白处，在家众会拜处悬挂三日，以示赏罚。

第三十条 设主记一人，以会货泉谷粟出纳之数。凡谷匣收满，主记封记，不许擅开，违者量轻重议罚。如遇开支，主记不亲视，罚亦如之。钥匙皆主记收，遇开支则渐次付之，支讫，复还主记。

The family should select an accountant who is responsible for the income and expenditure of a family. The accountant should check and count the grain yield and the goods carefully. When the barn is full, he should seal it with a seal number. It is not permittable to open it, otherwise the punishment will be imposed according to the situation. In case of expenditure, the accountant should handle and supervise it in person. The keys are at the accountant's disposal. In the event of drawing the grain, people should draw it in order according to the barn number. After drawing, they should return the keys to the accountant.

释义：设主记一人，以记载钱财谷粟出入之数。凡谷柜收满，即由主记负责贴上封条，做上记号，任何人不许擅开谷柜，违者根据其情节轻重议罚。如遇需要开支，主记不亲自监督也要议罚。谷柜钥匙皆由主记收藏，需要开柜支取粮食则按柜仓顺序依次领取钥匙，粮食支付完毕后钥匙再交还主记。

第三十一条 选老成有知虑者通掌门户之事。输纳赋租，皆禀家长而行。至于山林陂池防范之务，与夫增拓田业之勤，计会财息之任，亦并属之。

Choose someone who is sophisticate and knowledgeable to manage the family. He should inform the patriarch of the affairs as paying the taxes and renting the property. The prevention of forest fire and theft, the repair of pond dykes and the enlargement of a family's field as well as calculating the interests fall into the responsibilities of the housekeeper.

释义：选老成有见识能深谋远虑的子弟负责通掌门户之事。收租缴税等事，都要先禀告家长然后进行。至于用心防备山林火、盗和修整池塘堤坝，以及那些尽力增购田地开拓产业、计算钱财收取利息的责任，一并交给通掌门户者担任。

第三十二条 立家之道，不可过刚，不可过柔，须适厥中。凡子弟，当随掌门户者轮去州邑练达世故，庶无懵暗不谙事机之患。若年过七十者，当自保绥，不宜轻出。

The way of governing a family well is not to be too tough, nor to be too weak, but to be flexible and appropriate. All descendants should take turns to follow the housekeeper to the surrounding states and counties to broaden their vision and experiences, so as to avoid the chance of making mistakes and being confused. If the person is over 70 years old, he should not go out casually and should be careful of his own health.

释义：治理好一个家庭的办法，处事不可过于强硬，也不可过于柔弱，须刚柔适当。所有子弟都应轮流跟随通掌门户者去州县增长阅历，熟悉人情世故，以避免遇事糊涂不懂得办事的时机而造成的祸患。年过七十岁，应当自己保持安好，不宜轻易出门。

第三十三条 增拓产业，长上必须与掌门户者详其物与价等，然后行之。或掌门户者他出，必俟其归，方可交易。然又预使子弟亲去看视肥瘠及见在文凭无差，切不可鲁莽，以为子孙之害。

If the descendants want to add or develop new fields and industries, the patriarch and the keeper of the family must know the other party's physical objects and prices in detail. If the housekeeper goes out, they should wait until he comes back to close the deal. The patriarch should also send other family members to check the situation and the fertility of the field as well as the coupon to make sure nothing is going to be wrong. They should not act in a harsh and leave troubles to their offspring.

释义：增加和开拓新的产业，尊长与掌门户者必须详细了解对方的实物和价格，然后再行交易。如掌门户者外出，必须等待他回来，才可以进行交易。但是还必须预先派子弟到现场去察看田地的肥沃瘠薄情况，看清对方手中的券契有无差错。切不可鲁莽行事，给子孙留下祸害。

第三十四条 凡置产业，即时书于《受产簿》中，不许过于次日，仍用招人佃种。其或失时不行，家长朔望检点议罚。

The purchase of an estate must be immediately recorded in the *Newly-Bought Property Book*, ordinarily not later than the second day. In the meanwhile, they should rent them or hire people to cultivate them. If they did not do these, the patriarch should raise the issue on the family meeting on the 1st and the 15th day and discuss how to punish them.

释义：凡置办产业，应立即写在《受产簿》中，不许迟过第二日。同时马上招人租种。如有失时不办，家长初一、十五日检点时议罚。

第三十五条 增拓产业，彼则出于不得已，吾则欲为子孙悠久之计，当体究果直几缗，尽数还足。不可与驵侩交谋，潜萌侵人利己之心，否则天道好还，纵得之，必失之矣。交券务极分明，不可以物货遗负相准。或有欠者，后当索偿，又不可以秋税暗附他人之籍，使人倍输官府，积祸非轻。

When we increase the property, the other party sell it out of last resort. And we buy it because we want to think about our children and grandchild in the long run. So we should estimate the actual value of the property and pay the exact amount of money. Do not take the opportunity to undermine other's interest by deliberately colluding with the broker to drive down the land prices. Otherwise, Heaven is very fastidious about the retribution. Even if you get the property for the time being, there will be a time to lose it. Coupon deed and money must be delivered clearly on the spot, and it is forbidden to offset the delinquent land tax at the time of transaction. If there is a situation in arrears of land tax, we must collect it in time. We mustn't secretly spread the share of grain tax to other people's farmland and let them double pay to the government. This is really evil.

释义：增加产业时，对方是出于不得已而变卖的，我们要为子孙长久之计着想，当设身处地地估计产业实际的价值，尽数还足价钱。不可趁机暗地里萌发损人利己之心，与经纪人合谋故意压低地价。否则，老天是非常讲究报应的。即使你暂时得到田产，必然会有失去之时。券契和款项必须当场交付分明，不可在交易时抵消所购之田拖欠的田赋。如果有拖欠田赋的情况，过后必须及时讨取，但不可将粮税份额暗中摊附他人田籍中，让人家加倍地向官府缴税，这样做积祸是非常严重的。

第三十六条 每年之中，命二人掌管新事，所掌收放钱粟之类；又命二人掌管旧事，所掌冠婚丧祭及饮食之类。然皆以六月而代，务使劳逸适均。

Every year, two people are appointed to take charge of the "new things", that is, in charge of the affairs of collecting and telling money and grain; And two people are appointed to take charge of the "old things", that is, take charge of the food and drink, the weddings, the funerals as well as sacrificial rites. The old and new managers are to be replaced in the sixth month each year to make sure that people in a family share the hard work.

释义：每年任命二人掌管"新事"，即掌管钱财谷粟进出类事务；任命二人掌管"旧事"，即掌握戴冠礼、婚嫁、丧事、祭祀及饮食类事务。新旧管在每年六月份替代交换，必须让任事者能够均衡承担辛劳。

第三十七条 新旧管轮当，须视为切己之事。计会经理，自二十五岁至六十岁止。过此血气既衰，当优遇之，毋任以事。

People will take turns to be the new and the old manager but everyone should take it seriously as one's own business. People who are between 25 and 60 years old can be selected to handle the family accounting and managing. People who are over 60 years old should not be appointed to such affairs and should live a comfortable life for the loss of energy because of the senior age.

释义：新旧管在轮值之时，必须把掌管的事务看成自己切身利益之事。担任会计财务和经营管理的工作，自二十五岁至六十岁止。过了六十则血气已衰、精力不足，应当让其宽舒悠闲一点，不要担任此类事务。

第三十八条 新旧管皆置《日簿》，每日计其所入几何，所出几何，总结于后，十日一呈监视。果无私滥，则监视书其下，曰："体验无私"。后若显露，先责监视，次及新旧管。

The old and the new manager should set up a *Log Book* to record the daily income and expenditure and a summary is drawn at the end of the day. The log book should be presented to the supervisor every ten days. If it is true that there is no arbitrary use, the supervisor will sign with the word "experience selflessness". If excessive and improper expenses are found after that, firstly the supervisor should be punished and then the new and the old manager should be punished.

释义：新旧管都要设立《日簿》记载每日收入多少，支出多少，并汇总结果于后，每十天呈报给监视审核。如果确实没有胡乱使用，监视则签上"体验无私"四字，待后若发现有过度和不正当的费用开支，首先责罚监视，其次再责罚新旧管。

第三十九条 新管置一《总租簿》，明写一年逐色谷若干石，总计若干石，又新置田若干石。此是一定之额，却于当年十二月望日，以所收者与前谷总较之，便知实欠多少，以凭催索。后索到者，别书于《畸零簿》，至交代时，却入《总租簿》内通算。

The new manager would set up a *General Lease Book* to record in order the amount of grains of each type should be harvested, the total amount of grains should be harvested in a year and the amount in the newly-bought fields. On the 15th day of the 12th month, the manager should compare the amount of grains this year with that of last year. So they know the number in debt. They will record the later collected grains on other *Odd Books*. And these grains will be counted into the *General Lease Book* when handing over the position.

释义：新管设一本《总租簿》，有次序地写明本年全年应收租谷各个种类谷麦若干石，总计若干石，新置田应收租谷若干石。这是每年规定数额的租谷收入，但在当年十二月十五日，还要将所收租谷总数额与以前的数额逐一比较，便知道还欠多少，以便凭此催讨。以后催索所收到的，另外记于《畸零簿》上。至交代时再入《总租簿》进行汇总计算。

第四十条 新管所收谷麦，每匣收讫，即结总数报于主记。置《租赋簿》，令其亲书"某号匣系某人于某年月日收何等谷麦若干石"。量出之时，亦须置簿，书写"某匣舂磨自某日支起至某日用毕"，以凭稽考。

After each barn is filled, the total number of the barns will be reported to the chief accountant by the new manager. And then he will record the number in the *Lease Book* and ask the new manager to write："Who collect which kind of grain how many tons in which barn." When drawing the grain from the barn, it is necessary to write "On which date the grains are ground up in this barn" for future check.

释义：新管所收租谷，每个谷柜装满后，就将柜内总数报给主记，记入《租赋簿》，并由新管亲笔写明"几号柜某人于某月日收何等谷或麦若干石"。支取谷麦时亦必须将量出之数记入账册，写明"某柜舂谷或磨麦若干，自某月开始至某日舂磨完毕"，以备查考。

第四十一条 新管所管谷麦，必当十分用心，及时收晒，免致蒸烂；收支明白，不至亏折；关防勤谨，不至透失。赏则及之，若有前弊，罚本年衣资棉线不给。如遇称收繁冗，则拨子弟分收之。

The newly-harvested grain should be handled very carefully and exposed to the sun and withdrawn in time to avoid being ruined. The income and expenditure should be clear to avoid deficit. The security of the barn should be handled carefully to avoid loss in case of theft. Reward should be granted to the new manager. In the event of a deficit or loss as previously mentioned, the new manager would be penalized by deducting the clothing expense this year. In case that the job of weighing and collecting grains is too busy, the manager could send other family members to help him.

释义：新管所管的谷麦，要十分用心，及时收晒，以免因潮湿发热闷烂。收入支付应清楚明了，不使亏空；要小心谨慎地管好谷柜的封条印记，不要使所藏谷麦失去监管而遭到损失。有赏励也要给新管。如果出现前面提到的亏空、损失等过失，则罚扣新管当年的衣资和棉线。如遇收付过秤工作繁忙，可派子弟前来分担。

第四十二条 佃人用钱货折租者，新管当逐项收贮，别附于簿，每日纳诸家长。至交代时通结大数，书于《总租簿》，云"收到佃家钱货若干，总记租谷若干"。如以禽畜之类准折者，则付与旧管，支钱入帐，不可与杂色钱同收。

If a tenant pays the rent with cash or other goods, the new manager should collect item by item, recording it clearly and attaching it to the *Lease Book* and present the book to the patriarch for inspection on a daily basis. At the time of transferring duties, all such income figures are summarized and recorded in the *General Lease Book*, which states that "a number of money and goods have been received from tenants and a total number of grains have been received." If poultry animals are allowed to be converted into the rents, the livestock shall be handed over to the old manager for money and then recorded in the *Lease Book*. This should not be collected together with variegated money that tenant should give to the government.

释义：佃户用现钱或其他货物折价交纳租谷的，新管应当逐项分别收藏，将数目另外附在《租赋簿》中，每日交付家长检查。至交接职务时汇总所有此类收入数字，记在《总租簿》上，写明"收到佃户钱货若干，总计租谷若干"。如经允许以家禽牲畜折算成租谷交纳的，则将禽畜交给旧管，并从旧管处支钱记入《租赋簿》中，不可与佃户应交给政府的杂色钱一同收取。

第四十三条 田地有荒芜者，新管逐年招佃。或遇坍江冲决，亦即书簿，以俟开垦。开垦既毕，复入原簿，免致失於照管。

If there are barren fields, the new manager should hire tenants to grow grains on them. In the event of a flood that destroys a field, this should be recorded in the book immediately in preparation for the resumption of planting. If the field is resumed, write it into the original book to avoid loss of care.

释义：若有荒芜的田地，新管应当逐年招人租种。如遇大水冲毁田地，亦应立即将损失的田地记于簿册，以预备恢复种植。若恢复完毕，则在原簿上写明，以免失去照管。

第四十四条 田租既有定额，子孙不得别增数目。所有逋租亦不可起息，以重困里党之人。但务及时勤索，以免亏折。

Once the land rent is fixed, the offspring may not increase it under another denomination. If a tenant is in arrears with the land rent, his interest will not to be counted, thus aggravating the burden of the neighbors. However, arrears of the land rent must always be claimed in order to avoid losses.

释义：田租一经确定定额，子孙不得另立名目增加田租。佃户如有拖欠田租的，亦不可计其利息，从而加重邻里乡亲的困难。但拖欠的田租必须经常索讨，以免亏折。

第四十五条 佃家劳苦不可备陈，试与会计之，所获何尝补其所费。新管当矜怜痛悯，不可纵意过求，设使尔欲既遂，他人谓何。否则贻怒造物，家道弗延。除正租外，所有佃麦、佃鸡之类，断不可取。

The tenant's toil is hard to describe in words, the manager should calculate a year's income and expenditure for them to see which income of the harvest can make up for the cost! The new manager should be compassionate and should not indulge in over-seeking. He has to consider, if his desires are met, then what will others do? Otherwise, he will anger the Heaven, the creator of things, which will make it impossible to prosper the family business. Except for the provision of the field rent, no other form of rent can be charged.

释义：佃户的劳苦一言难尽，与他们算算一年的收支，其收获哪能弥补其花费！新管应当顾惜怜悯，不可纵意过求。必须考虑到，如果你的欲望满足了，别人将会怎么办？否则触怒创造万物的上天，就会使家业不能延长。除规定的田租之外，其余附加的佃麦佃鸡之类，坚决不可收取。

第四十六条 邻族分岁之饮，旧管于冬至后排日为之。

After the winter solstice, the old manager will arrange the annual banquet with the neighbors including the old and the young to ensure the harmonious neighborhood.

释义：旧管在冬至之后，逐日安排邻族一年一度长幼聚饮祝颂的宴席，以利于邻族和睦。

第四十七条 男女六十者，礼宜异膳。旧管尽心奉养，务在合宜。违者罚之。

Men and women who are over 60 years old will be served with special meals and treatment. The old manager should serve them with sincerity and try his best to meet their taste and requirement. The old manager who is against the rule will be punished.

释义：年满六十岁的男女族人，按照规定应该安排与众不同的膳食。旧管必须诚心奉养，想一切办法供给适合口味的食品。违者议罚。

第四十八条 新管簿书不分明者，不许交代。一应催督钱谷，须是先时逐项详注已未收索之数，于交代日分明条说，并承帐人交付。虽累更新管，要如出于一手，庶不使人欺隐。旧管簿书不分明者，亦不许交代。

If the accounting books are not clearly recorded, the new manager must not transfer the job. All that should be collected must be those recorded on the accounting books in detail, demonstrating in detail what has been collected and what has not been collected. The accounting books should be transferred to the new manager in time. Although the transferring of the manager occurs often, the accounting books should be clearly recorded as if they were recorded by one person. Thus, we can avoid deception and concealment by others. If the old manager did not record the accounting books clearly, he cannot transfer his post.

释义：新管账簿记载不清的，不得办理移交手续。一切应该催讨的钱谷，必须是以前逐项详细注清的已收和未收数字，在移交时，分条说明情况，与承接账目者一起交付。即使新管职务经常更换，但其账簿上的账目必须如出一人之手，这样才可以不受别人欺骗和隐瞒。旧管账目不记载清楚的，也不得办理移交手续。

第四十九条 所用监视及新旧管，其有才干优长、不可遽代者，听众人举留。

The appointed supervisors and the old and new managers should not be replaced in haste, especially those who are capable and excellent. Those capable supervisors and managers should be retained on the post based on the views of family members.

释义：所任用的监视和新旧管，如确有才干优长的，不可仓促更换，应该根据家众意见推举留任。

第五十条　设羞服长一人，专掌男女衣资事。宜先措置，夏衣之给，须在四月；冬衣之给，须在九月。不得临时猝办，如或过时不给，家长罚之。初生男女，周岁则给。

Set a post as Clothing Chief, managing in the production of men and women's clothing. He should plan the cloth-making in advance. The provision of summer clothing should be arranged in the fourth month and the provision of the winter clothing should be in the ninth month. If he cannot provide the clothes on time, penalty will fall on him by the patriarch. The newly-born babies will be given the clothing from one year old.

释义：设羞服长一人，专管男女服装制作之事。制作服装一定要预先安排，夏天服装在四月份供给，冬天服装在九月份供给，不得临时在计划之外置办。如不按时供给，家长要责罚羞服长。婴儿则满周岁开始供给。

第五十一条　男子衣资，一年一给；十岁以上者半其给，给以布；十六岁以上者全其给，兼以帛；四十岁以上者优其给，给以帛。仍皆给裁制之费。若年至二十者，当给礼衣一袭。巾履则一年一更。

Men will be given one set of clothing each year; Those who are over ten years old will be given half of that of the adults and their clothes are made of cotton; Those over sixteen years old will be given the same amount as adults with cotton clothes and silk clothes; Those who are over forty years old will be treated with preference for their clothes are all made of silk. In the meantime, they should be given extra money as tailoring fee; Those who are twenty years old should be given a set of formal dress and their hood and shoes should be renewed once a year.

释义：男子衣服一年供给一套。十岁以上按成年人定量减半供给，给以棉布；十六岁以上按成年人定量供给，用棉布兼丝绸；四十岁以上的给以优待，

全部用丝绸。同时发给裁制服装的费用。年龄二十岁的提供礼服一套。头巾和鞋一年更换一次。

第五十二条　妇人衣资，照依前数，两年一给之。女子及笄者，给银首饰一副。

Women's clothing is supplied every two years on the basis of the preceding example. Women who are over fifteen years old(Ancient women were allowed to get married after fifteen years old.)will be given a set of silver earrings, necklaces, rings and bracelets.

释义：妇女服装依据前例，每两年供给一次。女子到十五岁婚嫁年龄的，给银制的耳环、项链、戒指、手镯各一套。

第五十三条　每岁羞服长除给男女衣资外，更于四时祭后一日，俵散诸妇履材及油泽、脂粉、针花之属。

In addition to the clothes allocated to each family every year, on the day after the seasonal sacrifice, the Clothing Chief should distribute materials such as shoe materials, rouge and powder, needles and thread to women.

释义：羞服长在每年拨给家众衣资之外，在四时祭后一日，按份散发给诸妇做鞋材料、胭脂、水粉及绣花针线一类的用品。

第五十四条　各房染段，羞服长斟酌为之，仍置簿书之，毋使多寡不均。

The Clothing Chief should check the number of cloth to be dyed and stained and he should record it on the book to ensure even distribution.

释义：各房需染布匹绸缎的数目，羞服长须认真核查，并记入账册，以免多少不均。

第五十五条　子孙须令饱暖，方能保全义气。当令廉谨有为者以掌羞服之事，务要合宜，而无不足之叹。

Children and grandchildren must be fed adequately to preserve their loyalty. We should select those who are honest, prudent and capable to shoulder the responsibilities of Clothing Chief who can ensure adequate food and clothing and not to let people feel inadequate.

释义：子孙生活必须达到暖饱水平，才能保全义气。应该挑选廉洁不贪、谨慎小心并且能干有为的子弟担任羞服长之职。衣食务必要保证温饱，不要使人感到饱暖不足。

第五十六条　设掌膳二人，以供家众膳食之事，务要及时烹爨，不许干预旧管杂役，亦须一年一轮。

Appoint two persons as "Zhangshan" (Chef), who are in charge of the meals of the whole family, providing timely food and drink. They are not allowed to interfere with the responsibilities of the Old Manager and the other staff, with one-year tenure, upon the expiring date the position must be reelected.

释义：设"掌膳"二人，管理家众膳食，及时炒菜烧饭，但不许干预旧管和其他人员的职责，任期一年，到期必须更换。

第五十七条　择廉谨子弟二人，收掌钱货。所出所入，皆明白附簿。或有折陷者，勒其本房衣资首饰补还公堂。

Select two honest and prudent offspring of integrity to manage the finance. Every income and expenditure must be kept on the book with precision, if there is any loss or missing account, it must be compensated with properties like clothing and jewelry from their own family.

释义：选择廉洁奉公、工作小心谨慎子弟二人，掌管财务。钱财支出和收入必须准确无误地记入账册，若发现亏损和漏账的，则强制其以本房的衣资和首饰补还公堂。

第五十八条　择廉干子弟二人，以掌营运之事。岁终会算，统计其数，呈于家长。监视严加关防，察其私滥。

Select two honest and capable offspring to manage the business operation. In the year end, report the consolidated data to the patriarch. The daily work is to be supervised and closely examined to detect any wrong-doings like appropriating and wasting public funds.

释义：选择廉洁不贪、做事能干的子弟二人，掌管商业经营事务。年终会算时，将汇总数据呈报家长。平时监视严加审核，明察其有无私入腰包和浪费钱财等不规之行为。

第五十九条　子孙以理财为务者，若沉迷酒色、妄肆费用以致亏陷，家长覆实罪之，与私置私积者同。

Suppose the descendants in charge of finance are indulged in booze and women, or in reckless spendthrift, leading to operating losses and missing accounts. Once verified by the patriarch, they have to face the same punishment as those who purchase private properties.

释义：担任治理财务工作的子孙，如果有沉迷于酒色、胡乱开支、任意妄为，致使经营亏损和漏账的，经家长核实后，其责罚与私置产业私积钱财之罪相同。

第六十条 委人启肆，皆公堂给本与之，一年一度，新管为之结算，其子钱纳诸公堂。

Entrust someone to set up stores, with the money granted by the family board. The New Manager settles the accounts annually and transfer the interest into the family board.

释义：委托人员开设铺子，由公堂拨给资本。一年一度由新管与其结算，将利钱交纳公堂。

第六十一条 畜牧树艺，当令一人专掌之。须置簿书写数目，以凭稽考。然须常加点检，务要增益。如或失时不办，本人本年衣资不给。

Assign one person to manage farming and stockbreeding. A special book is set to keep the number, so as to be ready for future clarification. He should count the number from time to time, and make sure that the number is on the rise, rather than on the decline. If he misses the time for reproduction, resulting in no increase on the number, then he should be deprived of the yearly supply of clothing.

释义：畜牧种植方面的事，应专设一人管理。平时必须用专门的簿册记载数目，以备查考，同时要经常加以检点，其数字必须增加，不得减少。如错过季节导致没有增长的，则停止供给本人本年衣资。

第六十二条 设知宾二人，接奉谈论、提督茶汤、点视床帐被褥，务要合宜。

Assign two persons as "Zhibin" (Guest Attendants) to receive and entertain the guests. They see to the food and accommodation (beds, quilts and mosquito curtains etc.) for the guests, so as to make sure that the guests feel comfortable during their stay.

释义：设"知宾"二人，接待侍奉来往客人，陪客聊天，提醒催促茶汤，晚上安排客人就寝的床铺、被褥、蚊帐等事，务必使宾客感到舒适。

第六十三条 亲宾往来，掌宾客者禀于家长，当以诚意延款，务合其宜。虽至亲，亦宜宿于外馆。

"Zhibin" should report the visit of relatives and guests to the patriarch and entertain the visitors sincerely to meet their needs. Though close relatives, they should be accommodated in hotels at night.

释义：亲戚宾客来往，知宾应及时禀告家长，并诚心诚意给予款待，务必适合客人的心意。虽是至亲，晚上亦宜在外面安排旅店就寝。

第六十四条 亲朋会聚若至十人，旧管不许于夜中设宴。时有小酌，亦不许至一更，昼则不拘。

If the number of relatives and friends adds up to ten, the Old Manager should not entertain the guests at night. Even though it is for a quick drink, it should not be later than 9 o'clock at night, whereas there is no limit in the daytime.

释义：亲朋好友会聚若达到十人，旧管不应在夜间设宴招待，就是小饮亦不许到晚上九点。白天则不作限制。

第六十五条 亲姻馈送，一年一度，非常吊庆则不拘。此切不可过奢，又不可视贫而加薄，视富而加厚。

Bestow gifts to clan members once a year, unless there are special occasions like weddings and funerals. Such occasions should not be too luxurious. Gifts should not be reduced given that the other persons are poor, nor should it be increased for the affluence of the other. Whether it is for the rich or for the poor, everyone should be treated equally.

释义：向亲族赠送礼物一年一次，但情况特殊的吊丧、婚嫁庆贺则不限于此。这类事情切不可过于奢侈，更不可视对方贫穷而减少礼物，视对方富裕而增加礼物，无论贫富都应平等视之。

第六十六条 子弟未冠者，学业未成，不听食肉，古有是法。非惟有资于勤苦，抑欲其识斋盐之味。

Those not crowned or not finished schooling should not eat meat willingly, which was already practiced in the ancient times. The descendants are not only brought up to be hardworking, but also to understand the simple and poor country life.

释义：子弟未行冠礼的，学业未成，不能随意食肉。古时候早有这样的做法。不单要培养勤劳吃苦精神，而且也让他们了解农家清贫的生活。

第六十七条 子弟未冠者不许以字行，不许以第称，庶几合于古人责成之意。

Those not crowned are not allowed to use *zi* (courtesy name) or *hao* (style name) , nor are they allowed to be called according to the seniority among brothers and sisters, so as to meet ancestors' intention to urge the fulfillment of study.

释义：未行成年礼的子弟不许用字号，不许以排行辈分相称，以求符合古人督促其完成学业之意。

第六十八条 子弟年十六以上，许行冠礼，须能暗记四书五经正文，讲说大义方可行之。否则，直至二十一岁。弟若先能，则先冠，以愧之。

When descendants are of 16 years old, they can be crowned as adults, as long as they can recite the main part of the Four Books and Five Classics and explain the main idea of these books. Otherwise, they will be crowned when they are 21 years old. If the younger brother can recite before his elder brother, the younger one can be crowned first to encourage the elder.

释义：子弟年满十六岁可以举行冠礼。但必须能背诵四书五经正文，并能讲说其中的道理。否则至二十一岁再行冠礼。弟弟若先能背诵，则比哥哥先行冠礼，以此激励兄长。

第六十九条 子弟当冠，须延有德之宾，庶可责以成人之道。其仪式尽遵《文公家礼》。

Upon the crowning ceremony, one respectable and prestigious guest will be invited to teach the norms of adults, the rites of which follow *Family Rites of Wengong*.

释义：子弟在加冠时，应当聘请一位德高望重的贵宾，以教导他行成人之道。其仪式遵照朱熹《家礼》。

第七十条 子弟已冠而习学者，每月十日一轮，挑背已记之书，及谱图、家范之类。初次不通，去巾一日；再次不通，则倍之；三次不通，则分纷如未冠时，通则复之。

For those crowned descendants who learn at the family school they will be required to recite the memorized books as well as *Family Tree and Family Norms* every ten days. If the person fails to recite for the first time, the headcloth will be removed for one day; if he fails for the second time, the headcloth will be removed for two days; if he fails for a third time, then the bun will be untied just like the uncrowned, and the person will not be re-crowned unless he can recite the books again.

释义：已行冠礼又在家塾中学习的子弟，每月十日一轮，挑选已读之书及《谱图》《家范》背诵。初次不能背诵的除去头巾一日，再次不能背诵的去头巾二日，三次不能背诵的则解开其在加冠时所束的发髻，与未行冠礼的人一样，直至能背诵后复戴其冠。

第七十一条 女子年及笄，母为选宾行礼，制辞字之。

When a girl is of 15 years old (*jiji*, holding the hair together, means coming-of-age), her mother should invite guests to celebrate the coming-of-age ceremony, point out the merits and demerits, encourage her to behave herself with the standard of a good daughter-in-law, and are allowed to talk about marriage.

释义：女子年满十五岁，母亲要为其延请贵宾行成人之礼，指明她的优缺点，勉励她按照将来做好媳妇的标准要求自己，并且可以议论婚嫁之事。

第七十二条 婚姻乃人道之本。亲迎醮啐奠雁授绥之礼，人多违之。今一去时俗之习，其仪式并遵《文公家礼》。

Marriage is the origin of human reproduction. There were elaborate rituals such as toasting, saluting, presenting the string when seeing off the son-in-law, which were not observed anymore. Now all the customs are removed, the ritual of marriage follows *Family Rites of Wengong*.

释义：婚姻是人们繁衍延生之本。迎亲时敬酒、上礼、女家送婿出门时授车绳等等烦琐礼节，人们都已不去举行了。如今索性除掉一切时俗习惯，其仪式遵照《文公家礼》。

第七十三条 婚嫁必须择温良有家法者，不可慕富贵以亏择配之义。其豪强、逆乱、世有恶疾者，毋得与议。

The marriage subjects must be someone gentle and kind, with good family education. Don't violate the original meaning of marriage just for the sake of wealth. Don't tie the knots with those families that tyrannize in the countryside, defy one's superiors and start a rebellion, or suffer from incurable disease.

释义：婚嫁对象必须选择性格温良，且有家教的家庭子女。不要因为羡慕富贵而违反选择婚配的本来意义。那种横行乡里、犯上作乱以及家族有难以医治疾病的家庭不要与其论议婚姻之事。

第七十四条 立嘉礼庄一所，拨田一千五百亩，世远逐增，别储其租，令廉干子弟掌之，专充婚嫁诸费。男女各以谷一百五十石为则。

Set a House of Honoring Rites and grant 1500 mu(100 hectare) good farming land from the family board, with a year-on-year increase basis. This part of grain should be set aside separately and managed by an honest and capable descendant, used mainly for the expenses of marriage, with a same standard for both men and women, 150 shi (900kg) of grain.

释义：设立嘉礼庄一座，公堂拨良田一千五百亩，并且以后要逐年增加。这部分租谷应另行储藏，派定廉洁能干的子弟掌管，专用于婚嫁费用。男婚女嫁用谷以一百五十石为标准。

第七十五条 娶媳须以嗣亲为重，不得享宾，不得用乐，违者罚之。入门四日，婿妇同往妇家，行谒见之礼。

The most important purpose of marriage is to reproduce, so neither lavish feasts nor musical troupes should be allowed at weddings, those who violated these practices shall be punished. Four days after the wedding, both the wife and husband pay a formal visit to the father-in-law.

释义：娶亲最重要的目的是繁衍子孙，所以在举行婚礼时不得大办酒席、不得雇用乐班，违者给予责罚。入门四日后，夫妻双双至岳父之家，举行拜见之礼。

第七十六条 娶妇三日，妇则见于祠堂，男则拜于中堂，行受家规之礼。先拜四拜，家长以家规授之，嘱其谨守勿失；复四拜而去。又以房匾授之，使其揭于房阃之外，以为出入观省，会茶而退。

On the third day after the wedding, the new daughter-in-law meet the elder members of the family in the Ancestral Hall, the man call on his parents in the Main Hall, following the rituals required in the *Regulations*. First, the new couple kowtow for four times; then the patriarch pass on the *Regulations* and instruct them to comply with it prudently in case of any negligence; after that the couple kowtow another four times and then they can leave. The patriarch bestow them a plaque, which will be hung above the door, so that they can see it every time they come in or go out. Finally, everyone gets seated to have tea together, and then they can withdraw.

释义：娶媳第三日，新媳妇至祠堂与长辈相见，男则拜见父母于中堂，行家规之礼。先拜四拜，家长传授家规，并嘱咐其谨慎遵守，勿得有失，然后再拜四拜退下。家长又授以房匾一块，高挂于新人房门之上，以便他们出入时都能看到。然后大家一起饮茶，再各自退下。

第七十七条 子孙当娶时，须用同身寸制深衣一袭，巾履各一事，仍令自藏，以备行礼之用。

When a descendant is ready to get married, prepare a tailored *shenyi* (formal dress), a piece of headcloth and a pair of shoes, which should be kept intact by himself for future use.

释义：子孙当娶亲时，要制合身的衣服一套，头巾一块，鞋一双。制好后让子弟自己保藏，以备日后行礼之用。

第七十八条 子孙有妻子者，不得更置侧室，以乱上下之分，违者责之。若年四十无子者，许置一人，不得与公堂坐。

When a descendant is married and has kids, he shall not marry a concubine, in case the family status might be violated. Those who have broken the rule shall be punished. If someone is of 40 years old and still doesn't have any kid, he is allowed to marry a concubine, but she cannot sit in the family board.

释义：子孙已有妻子和孩子的，不得另娶小妾，以扰乱上下名分，违者给予斥责。若年满四十仍然没有孩子的，可酌情娶一人为妾，但小妾不得进公堂与大家同坐。

第七十九条 女子议亲，须谋于众，其或父母于幼年妄自许人者，公堂不与妆奁。

When it's time to choose a husband for a girl, it must be discussed with other members of the family. If her parents had betrothed the daughter to someone at an early age, the family board will not prepare dowry.

释义：女儿议亲择婿，必须与家众商量。如父母在女儿幼年时妄自许配于他人的，公堂不与嫁妆。

第八十条 女适人者，若有外孙弥月之礼，惟首生者与之，余并不许，但令人以食味慰问之。

After the marriage of a daughter, and upon the first month after the birth of the grandson, the maiden family will hold a *manyue* (the completion of the first month) ceremony for the first grandson. The kids born afterwards will not go through this ceremony, but gifts like food might be sent as greetings.

释义：女儿出嫁后，若有外孙满月，只为第一胎行满月之礼。以后出生的一概不行此礼，但可派人送食品慰问。

第八十一条 甥婿初归，除公堂依礼与之，不得别有私与，诸亲并同。

Upon the first visit of the grandson and son-in-law, the family board will present gifts according to the rites, others cannot present anything in private, nor can any other relatives.

释义：外甥和女婿第一次上门，除公堂依礼节给礼物外，他人不得私与，其余亲戚也一样。

第八十二条 姻家初见，当以币帛为赞，不用银罤。他有馈者，此亦不受。

For the first meeting of the in-laws, appropriate gifts like money and silk can be presented, but no banquet is needed. Other forms of gifts will not be accepted.

释义：姻家第一次见面，应当以适量的钱币绸缎为见面礼，不必摆设宴席。如果另外还有礼物馈赠，亦不能接受。

第八十三条 丧礼久废，多惑于释老之说，今皆绝之。其仪式遵《文公家礼》。

The funeral rituals have been given up for long, most of the present practices are under the influence of Buddhism and Taoism. Now all these are abolished and the funeral rituals should follow the practices as required in *Family Rites of Wengong*.

释义：丧礼久已荒废，现在的做法很多是受到佛教和道家的学说的蛊惑，现在都将其予以废除。家族的丧礼仪式均遵照《文公家礼》中的规定。

第八十四条 子孙临丧，当务尽礼，不得惑于阴阳非礼拘忌，以乖大义。

When the descendants are at funerals, all practices shall follow that required in *Family Rites of Wengong*. No superstitious behaviors like yinyang, taboos will be accepted so as not be violated.

释义：子孙举办丧事，务须按照《文公家礼》的规定，不得蛊惑于阴阳、禁忌等不合礼仪的迷信之说，以免违反情理。

第八十五条 丧事不得用乐。服未阕者不得饮酒食肉，违者不孝。

No music is played at funerals. When the family is still in mourning, they should not eat meat and drink wine, any violation will be regarded as impiety.

释义：举行丧事不得用鼓乐。服丧未结束，不得饮酒食肉，违者按不孝论。

第八十六条 子孙器识可以出仕者，颇资勉之。既仕，须奉公勤政，毋蹈贪黩，以忝家法。任满交代，不可过于留恋；亦不宜恃贵自尊，以骄宗族。仍用一遵家范，违者以不孝论。

For those talented and capable of being government officials, the family board may aid financially and give encouragement. After they have become government officials, they should respect justice and abide by the laws, work hard on government affairs. They should not embezzle money and engage in corrupt practices to humiliate family and violate family regulations. After the term of office expires, they should leave the post and should not feel reluctant to leave, nor can they feel honored and bloated with pride in front of clansmen. Though being out as government officials, they should still abide by *Family Regulations*. Any violation will be regarded as impiety.

释义：对有才能可以出仕的子孙，公堂应给以相当的资助和勉励。子孙出仕为官后，应该奉公守法，努力政事，不要涉足贪污受贿之事，以辱没家庭、触犯家法。任满离职，不要过于留恋官位，亦不应该自认为尊贵，对族人趾高气扬。即使外出为官亦必须遵守家族规范。违者以不孝论。

第八十七条 子孙倘有出仕者，当夙夜切切以报国为务。恤恤下民，实如慈母之保赤子；有申理者，哀矜恳恻，务得其情，毋行苛虐。又不可一毫妄取于民。若在任衣食不能给者，公堂资而勉之；其或廪禄有余，亦当纳之公堂，不可私于妻孥，竞为华丽之饰，以起不平之心。违者天实临之。

Those being out as government officials should always bear in mind how to repay the state, understand and sympathize with the poor and take care of them just like a mother. They should hold a heart of compassion for those with wrong charges, find out the truth and shall not abuse them. They should not take anything from the people. If the descendants are not self-sufficient in food and cloth, the family board can support them with funds. If the official's salary exceeds the cost of food and cloth, the surplus should be taken into the family board and never give it to his wife and children. In case they will buy lavish clothing to keep up with the Joneses, leading to resentment among others. Real misfortune will befall those who have violated.

释义：出仕为官的子弟务必早晚都要记住如何报答国家，关怀体恤穷困的黎民百姓，对他们应该如慈母爱护自己的儿子一样。对鸣冤求助的百姓要有哀悯恻隐之心，务必访查真情，不要苛刻虐待。更不能妄取百姓的一丝一毫。子弟在任时若衣食不能自给，公堂则给予资金补贴；俸禄若除衣食费用之外还有节余的，节余部分必须交纳给公堂，决不可私与妻子儿女，让她们竞相置办华丽的服饰，而使其他人产生不平之心。违者上天会实实在在地将不幸降到他们的头上。

第八十八条　子孙出仕，有以赃墨闻者，生则于《谱图》上削去其名，死则不许入祠堂。如被诬指者则不拘此。

If any descendant is infamous for bribe and embezzlement during his term in office, which is known by the family board, his name will be wiped out from the *Family Tree* if he is alive, and his memorial tablet will not enter the Ancestral Hall after his death. If the person is falsely charged, then he will not be restrained by this.

释义：子孙在出任官员期间，有因为贪污受贿而臭名远扬让公堂知晓者，生前则在《谱图》上削去其名字，死后则不许入祠堂。如被诬告冤枉者，则不拘于此。

第八十九条　宗人实共一气所生，彼病则吾病，彼辱则吾辱，理势然也。子孙当委曲庇覆，勿使失所，切不可恃势凌轹以忝厥其祖。更于缺食之际，揆其贫者，月给谷六斗，直至秋成住给。其不能婚嫁者，助之。

Clansmen come down in one continuous line. If they trap in trouble then we are in trouble too, if they are humiliated then we are humiliated, this is true regardless of reason and emotion. The descendants of Zheng should protect them whole-heartedly. Don't let them lose shelter because of poverty. Don't bully the weak which will disgrace the forefathers. When there is food shortage between two harvests, learn the needs of fellow villagers and provide them with 6 *dou* (60 liter) grain every month until the harvest in autumn. If there are some families fail to marry due to poverty, support them.

释义：同族之人，本来就是一脉所生的血缘兄弟。他们陷于困境就是我们陷于困境，他们遭受侮辱就是我们遭受侮辱，无论从道理上讲还是感情上讲都是这样。郑氏子弟应当对他们尽心保护，不要让他们因贫病而失去存身之地。切不可恃势欺压他们，以辱没上祖。在青黄不接的时候，应了解身处困境的乡邻，每月给谷六斗，供应到秋季收割时为止。如果有因穷困而不能娶亲或嫁女者，也给以帮助。

第九十条 为人之道，舍教其何以先？当营义方一区，以教宗族之子弟，免其束脩。

Are there any other better solutions than education for the rules of conducts? The family board will establish a family school to educate and encourage descendants benevolence, royalty and filial piety without any charges.

释义：为人之道，离开教育还有什么更好的办法呢？公堂将设立一所家塾，以仁爱忠孝教育和勉励子弟，并且免收他们的学费。

第九十一条 宗族无所归者，量拨房屋以居之。更劝勿用火葬，无地者听埋义冢之中。

When there are some homeless clansmen, provide them with shelter according to the actual situation. When someone passes away and doesn't have any land to bury in, persuade family members not to burn the body but bury it in the graveyard supported by the board.

释义：同族中有无家可归的，根据实际情况拨房屋给其居住。有人去世则劝家属不要采取火葬，没有土地埋葬的则由他们自己决定埋于义助的坟地之中。

第九十二条 立义冢一所。乡邻死亡委无子孙者，与给椁椟埋之；其鳏寡孤独果无自存者，时赒给之。

The family board sets a tomb for the unclaimed body. If the neighboring villager passes away without any children left behind, the family board provides a coffin to bury it. For those widowed clansmen who can't live on their own, the family board gives financial help to them on time.

释义：公堂设立埋葬无主尸骨的义冢一处。身后无子孙的邻里乡亲死亡，公堂提供棺材安葬。无法生存的鳏寡孤独族人，由公堂按时给予接济。

第九十三条 宗人无子，实坠厥祀，当择亲近者为继立之，更少资之。

Clansman with no children might break the sacrifice to ancestors, they should choose a close nephew to be stepson, with fund provided by the board.

释义：族人无子，有可能断绝祭祀香火的，应当选择近亲子侄立为继子，并给以资助。

第九十四条　宗人若寒，深当悯恻。其果无衾与絮者，子孙当量力而资助之。

If clansmen are extremely poor, be compassionate with them. If they have no quilts and padded coats in severe winter, fund them accordingly.

释义：族人如果非常贫困，应当对他们有怜悯恻隐之心。寒冬腊月他们如果确实没有被子和棉衣的，子弟应当量力给以资助。

第九十五条　祖父所建义祠，奉宗族之无后者。立春祭先祖毕，当令子孙设馔祭之，更为修理，毋致隳坏。

The public temple built by forefathers is to enshrine the tablets for those with no children. After the sacrificial ceremony for ancestors on the Spring Day, assign descendants to prepare food for memorial and do some repairs for fear of damage.

释义：祖辈所建的义祠，是用来供奉宗族中无子孙者牌位的。在立春日祭先祖仪式完毕后，应该派子弟设膳食给以祭奠，再为其修理，不致毁坏。

第九十六条　立春当行会族之礼，不问亲疏，户延一人，食品以三进为节。

On the Spring Day, the entire clan holds a sacrificial ceremony for ancestors together. Regardless of the closeness of kinship, it is customary to assign one person in each household to propose a toast for three times.

释义：每年立春日，当举行合族祭祀祖先的仪式，所有族人不问亲疏，每户一人，会聚时以进三次酒为适宜。

第九十七条　里党或有缺食，裁量出谷借之，后催元谷归还，勿收其息。其产子之家，给助粥谷二斗五升。

If the neighborhood is in need of food, we can lend them some grain according to our own strength. After the harvest, the grain is to be returned without any interest. If the family has given birth to a baby, provide them some grain, 2 *dou* 5 *sheng*(25 liter).

释义：街坊邻里有缺食的，可根据我们自己的力量拨出稻谷借给他们。秋收后仍然以稻谷归还，勿收利息。如有生育孩子的家庭，则提供他们助粥谷二斗五升。

第九十八条 展药市一区，收贮药材。邻族疾病，其症彰彰可验，如疟痢痛疖之类，施药与之。更须诊察寒热虚实，不可慢易。此外不可妄与，恐致误人。

The family board set a medicinal sector to collect herbs. When the neighboring kinsmen get sick, the symptoms of which are self-evident, like diarrhea, inflammation, or skin disease, give them medication timely. Upon diagnosis, observe the patients closely for the spells of fever and chills, deficiency and excess, don't overlook all the details and never abuse drugs, in case of misdiagnosing the illness.

释义：公堂开设药市一区，收贮药材。邻居亲族如有疾病，其症明显可以查看的，如疟疾、炎症、皮肤病之类，及时与之施药。在诊断时，须仔细察看病人的寒热虚实，不可轻心忽视，更不可乱施药物，以误他人。

第九十九条 桥圮路淖，子孙倘有余资，当助修治，以便行客。或遇隆暑，又当于通衢设汤茗一二处，以济渴者。自六月朔至八月朔止。

When the bridge is broken or the road is muddy, descendants should help to pay for the repairs and maintenance so as to provide convenience for passers-by. In the hot summer, from the first day of the sixth month to the first day of the eighth month, set up one or two tea houses in the crossroads of the main roads for passers-by to quench their thirst.

释义：塌梁桥和烂泥路，子孙若有余资，当助款加以修理，以方便行人。炎热夏季，自六月初一起至八月初一日，在通衢要道路口，摆设热茶一二处，以解行人之渴。

第一百条 里党之痒痀疾痛，吾子孙当深念之。彼不自给，况望其馈遗我乎？但有一毫相赠，亦不可受，违者必受天殃。

Regardless of the severity of illness, descendants should be sincerely concerned about our neighbors. If they cannot support themselves, how can they present anything to us? If the neighbors are suffering from poverty or illness, don't accept any bestowal at all, anyone who violates this will be punished by heaven.

释义：邻里乡亲无论大小疾病，我子孙应当深深挂念。他们自己都不能自给，怎么能希望他馈送我们什么东西呢？遭受贫病困难的乡邻，即使他们有一毫赠送我们，亦绝不可接受，违者必遭天祸。

第一百零一条 拯救宗族里党一应等务，令监视置《推仁簿》逐项书之，岁终于家长前会算。其或沽名失实及执吝不肯支者，天必绝之。此吾拳拳真切之言，不可不谨，不可不慎。

Everything including saving the kinsmen and neighborhoods should be recorded in *Book of Benevolence* one by one, and check the accounts with the Patriarch in the year-end. If anyone overstates the case to angle for fame, or fail to finance the funding because of meanness, heaven will surely leave no door open for them. These are words from the bottom of our heart, everyone should not be too careful on this.

释义：拯救族亲以及乡邻的一切事务，令监视及时写入《推仁簿》上，并逐次写明，年终与家长汇总核算。如有为谋取名誉故意夸大失实的，以及执掌者因为吝惜而不肯支付有关资助款项的，苍天必定会绝其后路。这是我等真心之言，在这件事情上大家绝对不可不认真小心。

第一百零二条 子孙须恂恂孝友，实有义家气象。见兄长，坐必起立，行必以序，应对必以名，毋以尔我，诸妇并同。

Descendants should be in awe of their parents and be kind to brothers, which is the atmosphere for a family of true filial piety and righteousness. When the younger descendant meets the elder one, he should rise to his feet; when they walk together, follow the right order; when they converse, address the others honorifically, but not frivolous addressing like you and me. The womenfolk of the family should abide by the same rule.

释义：子孙必须奉父母以恭敬，待兄弟以和善，这样才确确实实有孝义之家的气象。子弟遇到兄长坐必起立，行走时大小有序，应对时要尊称对方，不要轻佻地用你我相称。家族妇女也应同样遵守这一条规定。

第一百零三条 子孙之于尊长，咸以正称，不许假名易姓。

The junior Zheng descendants should address the seniors with courtesy name, style name, or seniority in the clan, never address them with first names.

释义：郑氏子弟中的晚辈对于长辈都要以字号和辈分称呼，不许直呼姓名。

第一百零四条 兄弟相呼，各以其字冠于兄弟之上；伯叔之命侄亦然，侄子称伯叔，则以行称，继之以父；夫妻亦当以字行，诸妯娌姒相呼并同。

When brothers greet each other, they should add their names before brother, the same is true for uncles to greet their nephews; when nephews greet their uncles, they use the seniority rank before uncle, the same is true for couples and sister-in-laws to greet each other.

释义：兄弟互相称呼，应将对方名字冠于兄或弟前面，伯、叔称呼子侄也如此，子侄称呼伯、叔则在排行之后再加上伯父、叔父，夫妻相称亦应当根据排行，各妯娌互相称呼也相同。

第一百零五条 子侄虽年至六十者，亦不许与伯叔连坐，违者家长罚之，会膳不拘。

Even when the nephews are of 60 years old, they can still not sit as equals at the same table with their uncles, anyone who violates this will be punished by the Patriarch, with only one exception, that is, when they are at dinner table.

释义：子侄即使到六十岁，也不许与伯父叔父平起平坐，违者家长议罚，但在会膳时不受本条规定限制。

第一百零六条 卑幼不得抵抗尊长，一日之长皆是。其有出言不逊、制行悖戾者，姑诲之。诲之不悛者，则重箠之。

The junior and the young must not speak up against the senior, including those who are one-day elder. If someone makes some offensive remarks and breaks the ethical standards, then he/she will be admonished. If they refuse to correct their errors after admonition, they will be flogged severely.

释义：小辈和年幼者不得顶撞长辈，包括年长一日者也是这样。如有出言不谦虚恭敬的，且行为违反道德准则的，先给以教育。教育之后仍不悔改的，则用鞭子重重惩罚。

第一百零七条 子孙受长上诃责，不论是非，但当俯首默受，毋得分理。

When descendants are receiving admonition and criticism from the seniors, they should lower their heads and accept the exhortation silently without defending themselves, no matter they are justified or not.

释义：子弟受到尊长或上辈的训斥和批评，不论是非与否，都应低头默受，不得分辩顶撞。

第一百零八条 子孙固当竭力以奉尊长，为尊长者亦不可挟此自尊。攘拳奋袂，恣言秽语，使人无所容身，甚非教养之道。若其有过，反复喻戒之；甚不得已者，会众箠之，以示耻辱。

Though the descendants should be committed to the seniors, the seniors should not intimidate the descendants, taking their standing as seniors for granted. It is improper to educate by rolling up one's sleeves and raising one's fist, speaking fiercely and filthily, making no place to run to for others. If the descendants committed some wrong-doings, they should be warned by reasoning and correct immediately. Suppose they failed to repent again and again, they would be flogged in public for humiliation.

释义：子孙固然应当竭力侍奉尊长，为尊长的也不可以自恃尊长身份而挟制子孙。捋起袖子举起拳头，气势汹汹语言肮脏，使人没有退步，这种方法根本不符合教养之道。如果子弟有过失，应该讲道理警告其改正；实在屡教不改的，才可以当众用鞭子惩罚他，以示耻辱。

第一百零九条 子孙黎明闻钟即起。监视置《夙兴簿》，令各人亲书其名，然后就所业。或有托故不书者，议罚。

Descendants should get up immediately as soon as the bell rings. The supervisor keeps the *Early Rise Record*, everyone should sign their own names on it in person and then start to do their own duties. If someone fails to sign his name deliberately, then he will face punishment.

释义：子孙在黎明时听到钟声应立即起床。监视设立《夙兴簿》，令各人亲自签字具名，然后做各人应做的工作。如有借故未签名的议罚。

第一百十条 子孙饮食，幼者必后于长者。言语亦必有序伦，应对宾客，不得杂以俚俗方言。

When descendants are at the table, the juniors should eat after the seniors. When they talk, they should behave themselves. When they converse with guests, they should not mingle in vulgar dialect.

释义：子孙在饮食时，年幼者必须后于年长者。讲话亦必须得体，应对宾客时，不能杂入粗俗的地方话。

第一百一十一条 子孙不得谑浪败度、免巾徒跣。凡诸举动，不宜掉臂跳足以陷轻儇。见宾客亦当肃行祗揖，不可参差错乱。

Descendants should not joke but act according to rules. They must not leave the head and feet naked. At no times should they dance with joy, jump up and down, lest they will be considered to be frivolous. When they meet with guests, they should behave solemnly and respectfully, making a bow with hands folded in front(saluting) when they meet and bid farewell to each other, with established order.

释义：子弟不得开玩笑，行事不可违反规矩。平时不准光着头和双脚。凡各举动不宜手舞足蹈、连蹦带跳，以免使人感到轻浮。会见宾客当严肃恭敬，进退作揖不可参差错乱。

第一百一十二条 子孙不得目观非礼之书，其涉戏谑淫衰之语者，即焚毁之，妖幻符咒之属并同。

Descendants should not read inappropriate books, whenever teasing or filthy expressions are found in books, they should be burned down immediately. The same is true to books containing witchcraft and spells.

释义：子孙不得看不正当的书籍，凡涉及玩笑嘲弄、有淫衰下流话语的书本，立即烧毁。妖幻符咒之类书籍也一样必须处理掉。

第一百一十三条 子孙不得从事交结，以保助闾里为名而恣行己意，遂致轻冒刑宪，隳圮家业。故吾再三言之，切宜刻骨。

Descendants should not gang up in village and act recklessly in the name of protecting the townsmen, so much so that it violates the laws and decrees of the country and destroys the family property. Therefore I repeat again and again for you to keep them in mind.

释义：子孙不得在乡里拉帮结派、互相勾结，以保护或协助乡里为名任意妄为，以致触犯国家刑律和法令，导致家业毁坏。所以我再三言之，一定要刻骨铭心牢记。

第一百一十四条 子孙毋习吏胥，毋为僧道，毋狎屠竖，以坏乱心术。当时以"仁义"二字铭心镂骨，庶或有成。

Descendants should neither learn to be inferior petty officials, nor to be monks and Taoists, they should not be on intimate terms with butchers and servants to disturb their mind. They should always remember by heart these two words—benevolence and justice, only in this way can they get somewhere.

释义：子弟不准去学做下等的小吏，不准充当和尚、道士，不准亲近屠夫、仆人，以坏乱心术。而应当时时将仁义二字铭心刻骨，这样才可能会有所成就。

第一百一十五条　广储书籍，以惠子孙，不许假人，以致散逸。仍识卷首云："义门书籍，子孙是教；鬻及借人，兹为不孝。"

The family board sets up Family Library to collect books for the benefits of descendants. Books should not be lent to others lest they will be missing. Every book should be marked with the following words on its preface："Books of Yi family, teachings of descendants; if lent to others, it is impiety."

释义：公堂设立"书种堂"多多储藏书，以造福子孙。不许借人，以免散失。而且在每部书的卷首都要做上标记，文字为"义门书籍，子孙是教；鬻及借人，兹为不孝。"

第一百一十六条　延迎礼法之士，庶几有所观感，有所兴起。其于问学，资益非小。若哤词幼学之流，当稍款之，复逊辞以谢绝之。

Confucian scholars good at rites and rules are invited to be teachers, so that it's possible to enlighten students through learning and yield progress. Those teachers help a lot in solving doubts and giving instructions. Those who can only talk illogically and teach little kids to write would be treated lightly and then dismissed politely.

释义：聘请通晓礼仪法度的儒生为师，就有可能让学子通过学习有所启发，学业上有所进步。那样的老师对于解答疑问、讲授学业的帮助是不小的。那种只会言语杂乱之词、教授幼童习字描红的先生，可以先略微款待，然后婉言辞退他们。

第一百一十七条　小儿五岁者，每朔望参祠讲书，及忌日奉祭，可令学礼。入小学者当预四时祭祀。每日早膳后，亦随众到书斋祗揖。须值祠堂者及斋长举明，否则罚之；其母不督，亦罚之。

When descendants are 5 years old, they should attend the learning at the Ancestral Hall on the first and fifteenth day of each month. They should also learn the sacrificial rites on the day of memorial. After they enter into primary school, they should attend the sacrificial ceremony in the four seasons. Every day after the breakfast they follow the others to the study to bow respectfully. If they are on duty at the Ancestral Hall and cannot go to the study, they should explain to the keeper of the study, otherwise they will be punished. If their mother does not urge them to do this, the mother will also face punishment.

释义：子弟满五岁者，每月初一、十五日参与祠堂听书讲学，到了忌日奉祭之时也要前去学习礼仪。入小学者，应当参与四时祭祀。每日早饭后随众到书斋恭敬行揖。必须在祠堂值日而不能去书斋者要到书斋长处说明理由，否则给予责罚；如其母不督促亦议罚。

第一百一十八条 子孙自八岁入小学，十二岁出就外傅，十六岁入大学，聘致明师训饬。必以孝悌忠信为主，期抵于道。若年至二十一岁，其业无所就者，令习治家理财。向学有进者弗拘。

Descendants begin to learn Chinese characters and phonology since they are 8 years old. They go out to study at 12 and start to learn the teachings of morals since 16. Renowned scholars will be invited to teach filial piety, royalty and honesty, so that students can master the truth of human relationship. If they are already 21 years old and get nowhere on learning, they are required to learn family governance and financial management. If they have been making progress in academic learning, then they are not limited by this rule.

释义：子孙自八岁入学学习文字、音韵，十二岁外出就学，十六岁开始学习关于道德教化学说。必须聘请名师教导，学习内容以孝悌忠信为主，以期望掌握为人处世的道理。若年至二十一岁，还未能在学业上有所成就的，令其学习治家理财。学业一向有上进的不拘于此。

第一百一十九条 子孙年十二，于正月朔则出就外傅。见灯不许入中门，入者箠之。

When descendants are 12 years old, they go out to learn on the first day of the first lunar month. When they come back after the lights are on, and then they cannot go through the middle gate, if they go inside they will be lashed.

释义：子孙至十二岁，在正月初一就必须外出就学。回家时如已经上灯，则不许入中门，入则鞭打。

第一百二十条 子孙为学，须以孝义切切为务。若一向偏滞词章，深所不取。此实守家第一事，不可不慎。

The main content of learning for descendants must be that of filial piety and justice, if priority is given to words and discourses, then it is not desirable. This is really the first principle of keeping a family, which requires great caution.

释义：子孙读书学习，千万要以孝义为必须学习的重要内容，若一向偏重和滞留于词章，是十分不可取的。这确实是守家第一大事，不可不慎。

第一百二十一条　子孙年未二十五者，除棉衣用绢帛外，余皆衣布。除寒冻用蜡屦外，其余遇雨皆以麻屦。从事三十里内并须徒步。初到亲姻家者不拘。

If descendants are under 25 years old, all garments are made of cotton, except that cotton-padded coat can be made of silk fabrics. Cotton-padded shoes are for the severe winter, linen shoes are for the rainy days. When the destination is within 30 li(15 km), walk back and forth, not for the first visit to the father-in-law.

释义：子孙年未满二十五岁的，除棉衣用丝织物制作外，其余皆用棉布。除寒冬腊月穿棉鞋外，其余遇雨皆穿麻鞋。外出办事三十里路内必须步行，如初到岳父家不拘于此。

第一百二十二条　子孙年未三十者，酒不许入唇；壮者虽许少饮，亦不宜沉酗杯酌，喧呶鼓舞，不顾尊长，违者箠之。若奉延宾客，唯务诚悫，不必强人以酒。

Descendants under 30 years old are not allowed to drink alcohol. They can drink a little bit in the prime of life, but not get indulged in it, lest they make loud noise after they get drunken, showing no respect for the elders. Those violate the rule will be lashed. When entertaining the guests, treat them with honesty and do not urge them to drink.

释义：子弟年纪未满三十岁的不许饮酒，壮年后虽可少饮，亦不宜没有节制地沉醉于酒中，以致喝醉后喧哗吵闹，不顾尊长，违者鞭打。如招待宴请宾客，就必须以诚对待，不必强人饮酒。

第一百二十三条　子孙当以和待乡曲，宁我容人，毋使人容我。切不可先操忿人之心；若累相凌逼，进退不已者，当理直之。

Descendants should live in harmony with the neighborhoods. Accommodate ourselves to others, rather than the vice versa. Don't get enraged when dealing with matters; if the counterpart keeps aggressive, argue with him bold and straight.

释义：子孙应当和睦对待乡邻，宁可以我容纳别人，不可让他人来迁就我。处理事情切不可先有对别人感到忿怒的心情；对方如果老是咄咄逼人，不肯罢休时，则应当理直气壮与之论理。

第一百二十四条 秋成谷价廉平之际，籴五百石，别为储蓄；遇时缺食，依原价粜给乡邻之困乏者。

When the grain price is cheap on harvest season, purchase 500 shi（3000kg）for separate storage. Upon food shortage before harvest, sell the grain to the neighbors in need at the purchased prices.

释义：秋收时节谷价低廉之际，籴谷五百石，另行储藏。在青黄不接乡邻缺食时，依收购时的价格粜给生活困乏的乡邻。

第一百二十五条 子孙不得惑于邪说，溺于淫祀，以邀福于鬼神。

Descendants should not be bewitched by superstition and heresy for blessings. They should not get obsessed with messy sacrifices not in accordance with social rites.

释义：子孙不得蛊惑于迷信邪说，沉迷于乱七八糟不合礼制的祭祀，求福于鬼神。

第一百二十六条 子孙不得修造异端祠宇，妆塑土木形象。

Descendants should not build unorthodox temples, make clay or wooden ghost figures.

释义：子孙不得修造不符正统的祠宇，妆饰和塑造泥塑木雕的鬼神形象。

第一百二十七条 子孙处事接物，当务诚朴，不可置纤巧之物，务以悦人，以长华丽之习。

Descendants should be sincere and honest in handling with people and affairs. Don't prepare fine possessions to please others, breeding the habit of luxury.

释义：子孙处事及待人接物，应当诚恳朴实，不要设置细巧之物，取悦于人，以滋长华丽的习气。

第一百二十八条 子孙不得与人眩奇斗胜两不相下。彼以其奢，我以吾俭，吾何害哉！

Descendants should not show off new fancy objects or contend for the upper hand. Others may enjoy their own luxury, we have our own plain life, and what's the point comparing with them?

释义：子孙不得与人炫耀新奇比赛争胜，两不相让。他人有他人的奢侈，我们有我们的俭朴，这对我有什么妨害呢？

第一百二十九条　既称义门，进退皆务尽礼。不得引进娼优，讴词献妓，娱宾狎客，上累祖宗之嘉训，下教子孙以不善。甚非小失，违者家长箠之。

Though our family is regarded as the righteous family, then the manners must be in accordance with the rites and norms. Descendants must not beckon prostitutes and musical artists to entertain guests with decadent music and obscene acts. Such behaviors will not only betray the good teachings of forefathers, but also abet descendants to do bad things. This is far from minor mistakes, those violated will be lashed.

释义：既然我家被别人誉为义门，那么举止行动都必须符合礼仪规范。子弟不得招引妓女和歌舞艺人，用靡靡之音和淫荡色相来供客人欢娱，因为这样做既辜负祖宗的良好教训，又等于教唆子孙行不善之事。此绝非小过失，违者家长鞭之。

第一百三十条　家业之成，难如升天，当以俭素是绳是准。唯酒器用银外，子孙不得别造，以败我家。

It is extremely difficult to achieve a successful family business, hence diligence and thrift will be the golden rules. Except the sacrificial vessels, silver is not allowed to make any other utensils, so as not to corrupt the family.

释义：成就一份家业，确实难于上青天，所以必须以勤俭朴素为准绳。除了祭祀用的酒器用银子制造外，不得用银子制造其他任何器具，以败我家。

第一百三十一条　俗乐之设，诲淫长奢，切不可令子孙听，复习肆之，违者家长箠之。

The folk music and theatre will always induce people to take in the customs of obscenity and luxury, which should warn descendants from entertaining. Those violated will be lashed.

释义：民间各种音乐戏曲，往往诱导人们沾染淫秽奢侈的习气，切不可让子孙反复去听、任意去做，违者家长鞭之。

第一百三十二条　棋枰、双陆、词曲、虫鸟之类，皆足以蛊心惑志，废事败家，子孙当一切弃绝之。

The game of chess and backgammon, music and pets befuddle one's mind and sap one's spirit by seeking pleasures until the decline of the family. Descendants should be prohibited to do anything of these.

释义：棋枰、双陆、词曲、虫鸟一类是迷惑人心的，使人玩物丧志，荒废正事甚至败坏家业，子弟应当全面禁绝这类事情。

第一百三十三条 子孙不得畜养飞鹰猎犬，专事侠游，亦不行恣情取厌，以败家事。违者以不孝论。

Descendants should neither raise eagles and hounds for entertainment, nor satisfy personal desires recklessly, leading to the decline of the family. Those violated are regarded as impiety.

释义：子孙不得畜养飞鹰猎犬，专求安乐游玩，亦不得恣意求取私欲，从而败坏家业。违者以不孝论。

第一百三十四条 吾家既以孝义表门，所习所行，无非积善之事。子孙皆当体此，不得妄肆威福，图胁人财，侵凌人产，以为祖宗积德之累，违者以不孝论。

Our family is viewed as a family of filial piety and righteousness. The morale and goal of our family is to do good to others. Descendants should understand this thoroughly and not do the following recklessly, bullying the neighborhoods, intimidating others' property, infringing upon others' business, lest they become a black sheep on the road of sowing virtues. Those violated are regarded as impiety.

释义：我家既然旌表为孝义家门，家族的风气和追求的目标，无非是要行积善之事。子孙都应当深刻领会，不得任意妄为，对乡亲作威作福，不得谋划胁迫他人钱财，侵犯他人产业，不要成为祖宗积德的败类。违者以不孝论。

第一百三十五条 子孙受人赞帛，纳之公堂，后与回礼。

Descendants should turn over the gifts from others to the family board, then the board send gifts in return.

释义：子孙接受别人送来的礼物，都应交纳公堂，然后由公堂给其回礼。

第一百三十六条 子孙不得无故设席，以致滥支。唯酒食是议，君子不取。

Descendants should not hold feasts for no reason, leading to excess expenses. A man of honor will not judge the moral standing by the abundance of the food and drink.

释义：子孙不得无故摆设宴席，造成过度的开支。仅以酒食是否丰盛来衡量人品，品格高尚的人是不会采纳这种做法的。

第一百三十七条 子孙不得私造饮馔，以徇口腹之欲，违者姑诲之；诲之不悛，则责之。产者、病者不拘。

Descendants should not prepare special food in private catering to the appetite for food. Those violated will be admonished first. If they refuse to repent, they will be disciplined. Puerperants and the sick are not limited.

释义：子弟不得私自开伙，以满足口腹的欲望。对违反者先给以教育，教育后仍然不悔改的，则给以批评指责。产妇、病人不限制。

第一百三十八条 凡遇生朝，父母舅姑存者，酒果三行；亡者则致恭祠堂，终日追慕。

Upon the birthday of one's parents and parents-in-law, he shall hold and make a toast with wine and fruits for three times if they are alive. Place offerings in the family temple and mourn for the whole day if they are dead.

释义：凡是遇到父母公婆生日，健在的则敬以酒馔果品三行。亡者则到祠堂敬上供品，终日追认思念。

第一百三十九条 寿辰既不设筵，所有袜履，亦不可受，徒蠹女工，无益于事。

No banquet will be held for the birthday and none of the gifted shoes and socks can be accepted, because this is just a waste of female needlework, and brings no benefits for health and longevity.

释义：寿辰既然不设筵席，所有送来的鞋袜亦均不可接受。因为那样不过是白白浪费女工，对于益寿延年没有益处。

第一百四十条 家中燕飨，男女不得互相献酬，庶几有别。若家长、舅姑礼宜馈食者非此。

If there is a banquet in the family, men and women should not make a toast to each other. There should be a prudent reserve between the genders. If it is between the patriarch and the parents-in-law, they can toast to each other according to the rites.

释义：家中设宴款待，男女不得互相敬酒回酒，应该男女有别。如果是家长和公婆，根据礼节可以敬酒的则不在此列。

第一百四十一条 各房用度杂物，公堂总买而均给之，不可私托邻族，越分竞买鲜巧之物，以起乖争。

The daily necessities needed will be purchased together by the family board and then distributed to each household. Never entrust neighbors to purchase novel and rare items beyond standards, causing unusual arguments.

释义：各房所用各种杂物，由公堂汇总购买后平均供给。不可私托邻居超越标准竞相购买鲜巧之物，而引起不正常的争论。

第一百四十二条 家众有疾，当痛念之，延良医以救疗之。

When the family member is ill, we should be sad and concerned, and invite doctors to treat the diseases in time.

释义：家众有疾病，应当痛惜挂念，及时延请良医急救治疗。

第一百四十三条 居室既多，守夜当轮用已娶子弟，终夜鸣磬以达旦，仍鸣小磬，周行居室者四次。所过之处，随手启闭门扃，务在谨严，以防偷窃。有故不在家者，次轮当者续之。

Assign one married descendant to guard the housing estate in turn. He shall beat the percussion instrument for alarm till dawn and patrol the estate for four rounds with the beat of chime stone. On the rounds, close the open doors and windows. Caution is the first priority in case of burglary. Anyone absent from home should make up for the patrol on the next round.

释义：居室既多，应该派已娶亲的子弟轮流守夜，自始至终敲击响器警戒直到天明，然后敲击小磬绕着居室巡查四次。在经过的地方，随手关好未关闭的门扇，重要的是认真严格，以防偷窃事故发生。有事外出不在家者下一个轮流时接上。

第一百四十四条 防虞之事，除守夜及就外傅者，别设一人，谨察风烛，扫拂灶尘。凡可以救灾之工具，常须增置，若篮油系索之属。更列水缸于房阁之外，冬月用草结盖，以护寒冻。复于空地造屋，安置薪炭。所有辟蚊蒿烬亦弃绝之。

Assign another man for the security of the night, excluding the night guardian and those out pursuing study. The main task is to closely check the candles and clear away the ashes on the hearth. The equipment to put out a fire should always be prepared, such as baskets, ropes etc. Besides, arrange a row of water vats outside the house in case of a fire. Put a straw lid on the top in winter to keep the vats from freezing and bursting. Build a special house for keeping firewood and charcoal on an open ground. Clear away all the ashes burnt for keeping off mosquitoes.

释义：夜晚安全防备之事，除守夜及出外就读的，另设一人，职责是严格仔细地察看火烛，扫拂灶上柴灰。凡是可以扑灭火灾的器具必须经常准备妥当，例如油篮绳索一类。另外在房门外要安放一排水缸以备救火之需，冬日用草结盖，以防寒冬冻裂。还要在空地上建造专用的房屋储存柴炭，所有驱蚊蒿所烧的余灰必须处理清楚。

第一百四十五条　旱暵之时，子弟不得吝惜陂塘之水，以妨灌注。

On time of severe drought, descendants should not skimp on the water in the ponds, hindering irrigation.

释义：在旱年大地干枯时，子弟不得吝惜池塘之水，以妨碍灌溉。

第一百四十六条　诸妇必须安详恭敬，奉舅姑以孝，事丈夫以礼，待娣姒以和。然无故不出中门，夜行以烛，无烛则止。如其淫狎，即宜屏放。若有妒忌长舌者，姑诲之；诲之不悛，则责之；责之不悛，则出之。

Every wife should be calm, respectful and stern, support parents-in-law with filial piety, attend up her husband with rites and treat sisters-in-law with gentleness. She should not go out of the inner yard without a good reason. When she walks around at night, she must light the road with a lamp. She is not allowed to go out without any light. If anyone is filthy and frivolous, she should be alienated. If anyone is jealousy of others, stirs up troubles by gossip, or alienates one person from another, she will be educated and disciplined first. If she refuses to repent, she will be reprimanded in public; and if she is still impenitent, then she will be driven away from the family.

释义：诸妇必须稳重、恭敬、严肃，供养公婆以孝顺，侍奉丈夫以礼节，对待妯娌以和顺。没有事情则不出内院门，夜里行走必须用烛光照明，没有烛光则不准外出。如果有妇人淫荡轻佻，就应该与她疏远。若有妒忌别人、拨弄是非、挑拨离间的，则先进行教育训导，经过教育仍不悔改的，则当众给以斥责，斥责后还是不改的，则赶出家门。

第一百四十七条　诸妇喋言无耻及干预阃外事者，当罚拜以愧之。

If any wife is nagging and frivolous, and even interferes with family affairs, she should be punished to kneel down and feel ashamed.

释义：诸妇言语啰唆、轻浮无耻，以及干预家族事务者，应当罚其跪拜，使其感到羞愧。

第一百四十八条 诸妇初来，何可便责以吾家之礼？限半年，皆要通晓家规大意。或有不教者，罚其夫。初来之妇，一月之外，许用便服。

As the wife is a new comer, how can we blame her not knowing our family rites? But she must understand the main idea of the family rules and regulations within six months. If she cannot understand after the time limit, then her husband will be punished. The newly-wed wife is allowed to wear casual dress after a month.

释义：新媳妇初来乍到，怎么能责怪她不懂我家的礼节呢？但是限她们在半年内必须通晓家规的大意。到期限仍不知晓的，则处罚其丈夫。刚刚入门的媳妇一个月后允许穿着便服。

第一百四十九条 诸妇服饰，毋事华靡，但务雅洁。违则罚之。更不许其饮酒，年过五十者勿拘。

The garment of wives cannot be too resplendent but must be modest and tidy, anyone violated will be punished. Wives are not allowed to drink alcohol, with the exception of those aged above fifty.

释义：媳妇的服饰不得追求华丽，但必须大方整洁，违者要受到处罚。更不允许媳妇饮酒，但年过五十岁的则不加限制。

第一百五十条 诸妇之家，贫富不同，所用器物，或有或无。家长量度给之，庶不致缺用。

Each wife may come with different dowry depending on the wealth of her maiden family, so each family may differ in the daily utensils. The patriarch should know all the facts and provide appropriate support to avoid shortage.

释义：各位媳妇因娘家贫富不同而陪嫁有多寡，故各家所用器物或有或无，家长了解情况后要给以适当供给，不使缺用。

第一百五十一条 诸妇主馈，十日一轮，年至六十者免之。新娶之妇，与假三月；三月之外，即当主馈。主馈之时，外则告于祠堂，内则会茶以闻于众。托故不至者，罚其夫。膳堂所有锁匙及器皿之类，主馈者次第交之。

The family meals are prepared by the wives of each family in turn every ten days, excluding those aged sixty. The newly-weds are granted with a three-month marital break, they need to take the shift for main chef immediately after the break.

As the main chef, she has to report to the family temple when she has to go out for purchase. The standards for meals are decided by family members during the tea break after the meal. If anyone refuses to take the chef task with excuses, her husband will be punished. The main chef should check the handover of the utensils and the keys for the dinner hall.

释义：家族膳食由各家主妇轮流主持，每十日轮到一次，年满六十岁者除外。新婚媳妇给婚假三个月，三月之后即应当参与主馈轮流值日。在担任主馈时，外出采买等事应禀告祠堂，膳食标准等事则在晚饭后会聚饮茶时听取家众意见。故意找借口推托不参与主馈工作的，处罚其丈夫。膳厅所有锁匙以及器皿，主馈应依次清点交接。

第一百五十二条 诸妇工作，当聚一处，机杼纺织，各尽所长，非但别其勤惰，且革其私。

All the wives should work together in one place, weaving and spinning, with each one showing her strong points. This can tell whether the person is diligent or lazy, and can get rid of the selfishness of certain persons.

释义：所有妇女务工劳作应当集中在一处，织布纺纱，各尽所长。这样既可以分出妇女何人勤快，何人懒惰，而且还可以革除一部分人的自私之心。

第一百五十三条 主母之尊，欲使家众悦服，不可使侧室为之，以乱尊卑。

The lady of the house is the elder respected by the whole family, the set of which is to convince the family completely. Never assign a concubine as the lady of the house, causing havoc with the superiors and inferiors.

释义：主母是家众所敬重的尊长，设立主母是要让家众心悦诚服，绝不可以小妾为主母，以乱尊卑。

第一百五十四条 每岁畜蚕，主母分给蚕种与诸妇，使之在房畜饲。待成熟时，却就蚕屋上箔，须令子弟直宿，以防风烛。所得之茧，当聚一处抽缫。更预先抄写各房所畜多寡之数，照什一之法赏之。

For the rearing of silkworms each year, the lady of the house distributes the silkworm eggs to each housewife, who rear the eggs in their own houses. When the worms are mature, put them on the worm-shelves in the worm-house. One descendant should be assigned to take the night shift and sleep in the worm-house for fear of a fire. The cocoons will be reeled into silk together. In addition, the number of silkworms kept should be recorded in advance, and rewarded with the one-tenth principle.

释义：每年养蚕，主母负责将蚕种分发给各家主妇，让她们各自在房中饲养。到成熟时，拿到蚕屋放在蚕箔上。这时应当安排一名子弟值更，宿于蚕屋，以防火烛。所得之茧应当集中缫丝。另外要事先记下各房养蚕数字，以什一之法给以奖励。

第一百五十五条　诸妇每岁所治丝绵之类，羞服长同主母称量付诸妇，共成段匹。羞服长复著其铢两于簿，主母则催督而成之。诸妇能自织造者，羞服长先用什一之法赏之，然后给散于众。

The silk and cotton produced by each housewife will be weighted by clothing chief(a middle-aged woman in charge of garment) and the lady of the house, and then given back to be woven into silk and cotton fabrics. Clothing chief keeps a record of the number of material handed out to each family, which is under the supervision of the lady of the house. If some housewives can already do the weaving, clothing chief can reward them with the one-tenth principle, and then hand out the silk and cotton to each family.

释义：各家主妇每年生产的蚕丝、木棉之类，由羞服长和主母过秤之后，交付给各家织成绸缎和布匹。羞服长将分发原料的数量记于簿册，主母则催督此事。各家主妇中有自己能够织造的，羞服长先以什一之法奖励，然后再将丝、棉分发给大家。

第一百五十六条　诸妇每岁公堂于九月俵散木棉，使成布匹。限以次年八月交收，通卖货物，以给一岁衣资之用。公堂不许侵使。或有故意制造不佳及不登数者，则准给本房。甚者住其衣资不给；病者不拘。有能依期而登数者，照什一之法赏之，其事并系羞服长主之。

The family board distributes cotton to each housewife to be woven into fabrics in the 9th month, which will be collected in the next 8th month and sold out to buy items for the garment expenses of next year. The family board cannot embezzle this fund. If someone produces defective goods on purpose or cannot finish the task on time, this will be translated to the fee of garment for this family. In serious cases, the fee of garment will be suspended, but persons of illness do not fall into this category. If this task can be finished with good quality on time, they will be rewarded by the one-tenth principle under the supervision of clothing chief.

释义：公堂每年在九月份给各家主妇分发棉花，织成布匹。限期在第二年八月份交收，全部出卖后购买物品，用于下一年的服装费用。这笔款项公堂不许挪用。如有人故意织成次品以及不按时完成的，则按照标准抵作本房的衣资，情节严重的将停发下一年衣资，但身体有病者不受本条规定约束。能够按时按数完成规定任务的，按照什一之法奖励，此事由羞服长主管。

第一百五十七条　诸妇育子，不得接受邻族鸡子彘胃之类，旧管日周给之。

When the housewife gives birth to a baby, they cannot accept the eggs and the pig's stomach presented from their neighbors. The old manager gives considerate supplies every day.

释义：各家妇女生育子女时，不得接受邻居赠送的鸡蛋、猪胃等食品。旧管应每日给予周到的供给。

第一百五十八条　诸妇育子，苟无大故，必亲乳之，不可置乳母，以饥人之子。

When the housewife gives birth to a baby, she should breast-feed the baby. A nannie is not allowed unless the puerperant passes away, so that other's baby would not fall in hunger.

释义：各家妇女生育孩子时，除非产妇去世，都必须亲自哺乳孩子，不可雇请乳母，而使他人的孩子陷于饥饿。

第一百五十九条　诸妇之于母家，二亲存者，礼得归宁。无者不许。其有庆吊势不可已者，但令人往。

If the wife wants to visit her parents' home, she can go back if they are still alive according to the rites. If the parents are not alive, then she could not go back. When there are some special occasions like weddings or funerals that she has to go, then she can entrust someone else to go on her behalf.

释义：各家妇女如果要回娘家，双亲健在则按照礼节可以回娘家看望父母；如父母已去世，则不准许。如娘家有喜庆或丧事之类不能不去的大事，只能委托他人前往。

第一百六十条　诸妇亲姻颇多，除本房至亲与相见外，余并不许。可相见者亦须子弟引导，方入中门，见灯不许。违者会众罚其夫。主母不拘。

The wife is only allowed to meet the closest relatives, the rest of which are not permitted. When her parents or brothers come, they can go into the middle gate only when guided by the descendant. When the night falls and the lights are on, they have to leave immediately. If someone breaks this rule, the patriarch will punish the husband in front of the whole family. The lady of the house is an exception.

释义：各房女家亲戚众多，除了本房最亲近的亲属允许相见外，其余一律不许相见。如父母、兄弟等允许相见的亲属亦必须由子弟引导，方可进入中门，但一到天黑掌灯即不许相见。违者家长会集家众罚其丈夫。但主母不受本条规定限制。

第一百六十一条 妇人亲族有为僧道者，不许往来。

If the wife has some relatives who are monks or Taoists, don't get in touch with them.

释义：各房妇女有做和尚、道士的亲族，则不许与他们往来。

第一百六十二条 朔望后一日，令诸孙聚揖之时，直说古《列女传》，使诸妇听之。

On the first and fifteenth day of each month, when the offspring meet to perform the rites, the stories recorded in *Stories of Famous Women* should be told accurately to educate the women of the family.

释义：每月初一、十五次日，在诸孙聚集会揖之时，要如实讲说古代《列女传》上记载的妇女事迹，让全家妇女接受教育。

第一百六十三条 世人生女，往往多致淹没。纵曰女子难嫁，荆钗布裙有何不可？诸妇违者议罚。

Common people might as well drown baby girls. Though girls are difficult to get married without high dowry, it is pretty good to marry with ordinary jewelry and plain cotton clothing. Any women violate this regulation will be punished.

释义：世人生育女孩，往往有将其溺毙的现象。即使说女子没有高额陪嫁就难以出嫁，难道用普通首饰、粗布衣裙做嫁妆有什么不可以？有妇女违反本条规定者议罚。

第一百六十四条 女子年及八岁者，不许随母到外家。余虽至亲之家，亦不许往, 违者重罚其母。

When girls are of eight years old, she cannot go to the grandparents' home with her mother, as well as other relatives' home, even though the closest ones. Upon violation, the mother will be severely punished.

释义: 女孩年满八岁, 不许随母亲到外祖家去, 其他的亲戚家即使是至亲亦不许前往, 违者重罚其母。

第一百六十五条 少母但可受自己子妇跪拜, 其余子弟不过长揖。诸妇亦同。有违之者, 监视议罚。死后忌日亦同。

A concubine can only accept the kowtow of her own children and daughter-in-law. Other descendants can greet with a long bow. The same is true between the housewives. Any violation will be punished by the supervisor(a monitor). After the death of the concubine, the same rite is required on the day of memorial.

释义: 少母只可以接受自己的亲生子女和儿媳跪拜, 其余子弟长揖即可。各妇亦相同, 有违反规定者由监视议罚。少母去世后忌日祭祀亦同。

第一百六十六条 男女不共围溷, 不共湢浴, 以谨其嫌。春冬则十日一浴, 夏秋不拘。

Men and women cannot use the same toilet and take a shower in the same bathroom to avoid arousing suspicion. Bath every ten days in spring and winter, there is no restriction in summer and autumn.

释义: 男女不得共用厕所, 不得同浴室洗澡, 以避其嫌疑。春冬两季十日一浴, 秋夏不限制。

第一百六十七条 男女不亲授受, 礼之常也。诸妇不得用刀镊工剃面。

It is the basic requirement of rites for men and women not to get close touch. Hence, women cannot have their faces shaved by barbers, so as to avoid close contact.

释义: 男女必须避免直接接触, 这是礼仪的基本要求。因此, 家族妇女不得让整容匠人剃面, 以免直接接触。

第一百六十八条 庄妇类多无识之人, 最能翻斗是非。若非高明, 鲜有不遭其聋瞽, 切不可纵其来往。岁时展贺, 亦不可令入房闼。

Rural women are usually ignorant, who are good at sowing discord. For those who are not very smart, it is quite often for them to be fooled and tricked by them.

As a result, women in the household should not be allowed to get in touch with them. Upon festive celebrations like Lanterns Festival, Mid-Autumn Festival etc., rural women are not allowed to get into the doors.

　　释义：乡村妇女一般都是无识见之人，最会拨弄是非。不是很高明的人，很少有人不会被她们愚弄欺骗，切不可允许她们与家中妇女互相往来。在每年元宵中秋等节日举行庆贺之时，亦不可让她们进入房门。

后 记

2017年10月，为迎庆党的十九大召开，展示金华丰厚的历史文化遗产，反映传统家规家训所倡导的忠孝、廉洁、礼仪等中华优秀传统美德，由金华市文物局主办，金华市博物馆承办的"古训新风——金华家规家训联展"正式开展。展览通过婺学理论引导，呈现出金华各名门望族以立规明训的传统家风，秉承中华传统文化，展现一代代金华人的立身处世、持家治业之道。展览的成功举办离不开市文物局、市博物馆领导和同仁的大力支持，也离不开兰溪、义乌、永康、东阳、武义、浦江、磐安等各县市博物馆的鼎力相助，在此一并向他们表示由衷的感谢。

本书以展览为脚本，在上海财经大学浙江学院曹艳梅、李慧等诸位老师的辛勤努力下，作了增补和完善，最终汇编为以介绍金华家规家训为核心内容的中英双文版科普类读物，以飨读者。

"欲治其国者，先齐其家。""家齐而后国治，国治而后天下平。"（《礼记·大学》）一个国家，需要一种精神力量的支撑，它可以是一种主流价值观，可以是一个民族的灵魂，也可以是一个家庭的家规和家训。这种精神力量自古便一脉相承，在家国同构的中国社会支撑着一代代中华儿女创造出悠久和灿烂的文明。

家规和家训多以谱牒为载体而得以传承，是家谱的重要组成部分，它植根于以宗法为纽带的中国农业社会。金华是农业文明起源地，早在距今约1万年的上山文化时期，就诞生了中国南方迄今最早的农业定居聚落。在人类仍主要依赖自然馈赠的蒙昧时期，上山文化各聚落内部以及聚落之间必然是通过一定精神意识形态，来共同维系当时的氏族社会关系。

从金华各地先秦、汉六朝、隋唐等时期出土的文物和保存至今的古迹中不难发现，金华人自古便重视农业的发展，也正因于此，这里的民德归于素朴，社会走向法治。南宋时，出身于"东莱吕氏"的吕祖谦，以史明理、测古证今、变化应用，兼取各家所长，创建"婺学"，形成独具地域特色的思想体系和处世哲学。金华家规家训深受"婺学"思想的浸润，更多地保留了社会伦理与个人操守的教育，以其独特的方式演绎着社会的变迁，成为当代人为人、持家、处世、治学、经商之道的规范和标准，也是"弘扬和践行新时代金华精神"的文化源泉。

金华市文物保护与考古研究所 文博馆员 徐峥晨

2021 年 11 月

Postscript

In Oct. 2017, Jinhua Administration of Cultural Heritage, together with Jinhua Museum, organized a formal exhibition "Ancient Norms with New Traditions—A Joint Exhibition of Family Instructions in Jinhua", which aims to reflect the excellent Chinese virtues advocated in traditional norms and regulations, so as to celebrate the opening of the 19th CPC National Congress and to showcase the abundant cultural legacy of Jinhua. Based on Wu-learning, the exhibition displayed the traditions and teachings that the noble families in Jinhua have guided to set rules and regulations, which have inherited the traditional Chinese culture and embodied the ways of conducting in society and governing the family and business for each generation of Jinhua. The success of the exhibition came with the joint efforts of Jinhua Administration of Cultural Heritage, Jinhua Museum, as well as the strong supports of the county museums in Lanxi, Yiwu, Yongkang, Dongyang, Wuyi, Pujiang and Pan'an. I'd like to express my heartfelt thanks to them.

Take the exhibition as the script, the teachers from Shanghai University of Finance and Economics Zhejiang College had supplemented and refined the manuscript and finally compiled it into a bilingual popular science reading, with the norms and traditions of Jinhua as the core of the book.

"Those who desire to govern the state must manage their families well first." "Only after the family is disciplined can they govern the state, and then govern the whole world" (*The Great Learning, Book of Rites*). A nation needs a kind of spiritual support, which can be a mainstream value, a soul of the nation, or a family's regulations and norms. This spirit has its origin in ancient China, which has empowered the Chinese descendants to create long-standing and brilliant civilization in Chinese society founded on the same structure as the clan.

Originated from the agricultural Chinese society bonded by the patriarchal clan system, family regulations and norms were mainly carried forward in the form of genealogical records and were an essential part of the family tree. Jinhua is the birthplace of agricultural civilization, where the earliest agricultural settlement in southern China came into being during the Shangshan Culture about ten thousand years ago. During the ignorant period when humans were mostly dependent on the bestowals of the nature, there must be some kind of ideological bonds between the groups within the settlement during Shangshan Culture, so as to keep the clan system in the society.

Judging from the relics unearthed from the pre-Qin, the Han Dynasty and the Tang Dynasty and the historic sites kept intact till today, it can be seen that Jinhua people have always emphasized agriculture, therefore, the mores of people in this land are simple and sincere, paving the way for a law-based society. In the Southern Song Dynasty, Lv Zuqian, born in "the Lvs from Donglai", created "Wu-learning" by incorporating the strong points of other schools of thought. He took history to tell right and wrong, made the past serve the present and adjusted to changing circumstances, making a whole set of ideological system and philosophy of life.

Well-nourished by "Wu-learning", Jinhua norms and regulations have kept a great deal of teachings on social ethics and personal conducts, showing the transformation of society in a unique way. As for the modern people, the norms have become the rules and regulations for personal conduct, family governance, social intercourse, pursuit of learning, and business undertaking, which has also proved to be a rich source of inspiration for carrying forward and living up to the spirit of Jinhua in the new era.

Xu Zhengchen

The Cultural Relics Protection and Archaeology Institute of Jinhua

Nov. 2021